The Economic Benefits
of
Forestry Research

The Economic Benefits of Forestry Research

William F. Hyde, David H. Newman, and Barry J. Seldon

Iowa State University Press / Ames

William F. Hyde is Chief, Water Branch, Economics Research Service, United States Department of Agriculture, Washington, D.C.

David H. Newman is Assistant Professor, D.B. Warnell School of Forest Resources, University of Georgia, Athens.

Barry J. Seldon is Associate Professor, School of Social Sciences, University of Texas at Dallas, Richardson.

Table 2.1 is reprinted with permission, from Vernon Ruttan, *Agricultural Research Policy* (Minneapolis: University of Minnesota Press, 1982), pp. 242-243.

Table 4.1 is reprinted, with permission, from Robert G. Anderson, *Regional Production and Distribution Patterns of the Plywood Industry* (Tacoma, Wash.: American Plywood Association, 1982), p. 18.

⊛ Printed on acid-free paper in the United States of America

First edition, 1992

Library of Congress Cataloging-in-Publication Data

Hyde, William F.
    The economic benefits of forestry research / William F. Hyde, David H. Newman, and Barry J. Seldon.—1st ed.
        p.   cm.
    Includes bibliographical references (p.   ) and index.
    ISBN 0-8138-0849-9
    1. Forest products industry—Research—United States.   2. Forests and forestry—Research—Economic aspects—United States.
    I. Newman, David H. II. Seldon, Barry J. III. Title.
    HD9755.H93 1992
    338.1'7498'0973—dc20   92-14835

Dedicated to

Dolly, Barbara, and Wendy,

for enduring support and encouragement

and to

T. Dudley Wallace,

for being our teacher, colleague, and friend

CONTENTS

Foreword   ix
  by John M. Antle
Preface   xi

Chapter 1:   **Introduction**   3

Chapter 2:   **A Short History of Technical Change**   23

Chapter 3:   **The Benefits of Forest Products Research:**   34
             A Dual Approach

Chapter 4:   **The Softwood Plywood Industry**   59

Chapter 5:   **Further Examinations in Forest Products Industries**   81

Chapter 6:   **Technical Change in the Southern Pine Industry**   114

Chapter 7:   **Distributive Effects of Technical Change in**   153
             Southern Softwood Forestry

Chapter 8:   **The Net Benefits of Southern Softwood**   173
             Forestry Research

Chapter 9:   **Summary, Conclusions, and Policy Implications**   196

Literature Cited   231
Index   241

FOREWORD

TECHNOLOGICAL INNOVATION has been the major source of the high rate of productivity growth in agriculture since the 1950s. Publicly funded agricultural research has played an important role in developing modern agricultural technology. Rates of return to public investment in agricultural research have generally been much higher than rates of return on investment in the private sector.

This important book is the first comprehensive assessment of public investment in the forestry sector—timber growth and management and forest products research—in the United States. The authors skillfully combine a rich data base for the 1950-1980 period with modern econometric techniques. They construct a convincing case that publicly funded forest products research was highly successful, whereas timber growth and management research has been less successful. This research benefitted the timber industry, but in the long term a large share of the economic benefits of this research accrued to consumers of forest products. Overall, the record shows that publicly funded forestry research in the United States, like other agricultural research, has played an important role in productivity growth since 1950.

There is a growing concern in agriculture, however, that high rates of productivity growth may not be sustainable in the US or elsewhere. Signs of stagnant or declining yields in the principal "green revolution" commodities, high rates of deforestation, resource depletion caused by population pressure, and other indicators of environmental stress have made sustainability one of the leading questions facing researchers. The importance of obtaining the greatest productivity from forest resources thus should be evident. The findings presented in this volume will be useful in evaluating future forestry research strategies, both in the United States and in the developing countries where the sustainability issue is a particular concern.

In the United States, where timber is abundant relative to other capital and labor, capital-intensive forest products research has been successful but there is little incentive to invest in or adopt improved timber management practices. Observing that in developing countries labor is often abundant relative to capital, the authors argue that there may be little opportunity to use many of the modern forest products technologies. In those tropical developing countries where forests are even more abundant than in developed countries—and where much of the concern over deforestation is focused—there is also likely to be little private incentive to adopt improved timber growth and management practices. Considering that there appears to be an abundant supply of timber to the world market for the foreseeable future, the unfavorable economics of improved forestry management practices are not likely to change soon. The lack of well-defined property rights for forests and perverse public policies further erode the incentive to adopt improved management practices.

These observations lead to an important hypothesis for future forestry research that needs to be addressed in detailed studies of forestry productivity in other countries. The authors conclude that improved forest management is not likely to contribute substantially to the solution of problems associated with tropical forests. They suggest that other areas of research appear to hold greater potential: adapting modern capital-intensive logging technologies designed for labor-scarce developed countries to the labor-abundant conditions in developing countries; understanding the legal and institutional barriers to the development of forest property rights; and understanding the impacts that other public policies have on the use of forestry resources and how those policies can be modified so as to mitigate unintended effects.

Answers to these and other questions regarding the appropriate future path for forestry research will no doubt require careful, detailed analysis based on data specific to each situation. The present study provides an analytical framework and empirical methods that should be useful in assessing the productivity of forestry research in other parts of the world.

JOHN M. ANTLE

PREFACE

THE PURPOSE of this book is to provide a careful assessment of previous US forestry research programs, with the more general objective of providing policy insights into (a) the distinguishing characteristics of research programs with strong prospects for producing a stream of future benefits and (b) the technical and market bottlenecks which deter even greater research successes. The individual research analyses in the book should also display general approaches useful for subsequent assessments of forestry research.

Our approach relies on the examination of several specific cases. We have the expectation that these cases provide insights into broader, more generalizable research programs. Our inquiry was important for US forestry researchers when we began in 1983 because US forestry research had seldom undergone technical inquiries of this kind and because US forestry research was embarking upon a period of unusually tight budgets. Tight budgets beg careful justification of all forestry research programs. Many developed countries share the experience of tight research budgets. Therefore, many developed country forestry institutions may also gain insight from our final US observations.

There is, however, an even greater need today, in 1992, for knowledge of both (a) the technical methods of research evaluation and (b) the more general lessons gained from reflections on previous US research experiences and their suggestions for the future. This greater need resides in the rapidly expanding forestry research programs for developing countries.

Forestry has become an exciting topic in economic development and an immense amount of money is being spent on all aspects of forestry, including forestry research, in developing countries today. Much of this money comes from international lending agencies that often require repayment regardless of the eventual success of any local project. Thus, while forestry research programs may promise great gains for the rural poor in many developing countries, they also promise extended indebtedness if these programs are chosen with less than the greatest care and best

xi

economic foresight.

We hope that our final chapter brings these public policy concerns together. We hope that it provides useful reflection on the US' greater and lesser forestry research successes of the past and intuition with regard to the US' forestry research future. We also hope that it provides justifiable extrapolations to the more general forestry research situations occurring around the world, regardless of the level of local economic development.

In our opinions, this book makes three important contributions:

• Chapters 3-5 make both a technical and an empirical contribution. They introduce the "dual" to the general research evaluation literature. This production/supply dual enables derivation of both average and marginal research impacts from the same empirical economic specification. Furthermore, the softwood plywood and sawmill cases in chapters 4 and 5 also obtain results with a degree of statistical reliability unusual in the general research evaluation literature—and sufficient to provide a most confident general understanding of the markets facing these two industries.

• Chapters 6-8 display the true difficulties of understanding aggregate measures of timber growth and management over time, yet they also derive confident intertemporal results for the dynamic and multiple product southern pine sector of the US forest industry. These three chapters offer a solid point of modeling departure for future analyses of aggregate timber production.

• Chapters 1, 2 and 9 first anticipate, then reflect upon, the great disparities in the historical research payoffs between forest products research and timber growth and management research in the US. More importantly, they rely on general empirical economic knowledge to reflect on why these disparities (while previously unexamined and often unanticipated) should not be surprising. Furthermore, chapter 9 discusses why the performance disparities between forest products research and timber growth and management research have strong implications for the efficiency of future forest research investments in the US—and in other countries with even less-developed forestry sectors. This should be critical insight for discussions of future research budget allocations.

Our intention throughout this book is to produce reliable economic assessments. This implies considerable technical economic detail in the middle six chapters. For this reason, we have tried to produce introductory and concluding chapters with sufficient background and summary material to ease the understanding of the more technical chapters. Furthermore, the technical chapters themselves include lengthy introductions and summaries that should further ease the reader's way through the more detailed technical discussions.

Graduate students of forest economics and policy should be able to read the entire manuscript without great difficulty. Chapter 9 is a survey chapter written to stand on its own. It should be accessible for those interested policy analysts with severe personal time constraints. If the latter individuals choose to examine the underlying analytical principles and assumptions without going into technical detail—then the introductions and conclusions to chapters 3-8 should be satisfactory.

Many have helped us through the long process of preparing this book. None have been so effective as Bob Buckman and Dudley Wallace. Bob perceived the usefulness of research evaluation in forestry before we did ourselves. He encouraged honest research results without flinching and he added to our own education with his terse but penetrating insights into the proper relationship between public research administration and research results with diverse, and even challenging, policy implications. Indeed, Bob challenges some of our conclusions but his encouragement has never faltered. Our first preference is to dedicate the book to Bob, along with Dudley. That we do not only reflects our respect for his independent, and sometimes different, opinion about our results. Bob is our model as an administrator of public research.

Dudley provided the senior faculty support necessary to build and maintain our academic program in resource economics at Duke. His unsurpassed insistence on combining rigorous applications of economic theory with good statistics helped make the three of us adequate economists (we hope) as well as "furresters."

The US Forest Service and the Cooperative State Research Service funded the largest shares of our work. George Dutrow of the Forest Service, and later Duke, and Sam Gingrich, then of the Forest Service, now retired, thrust us into the topic and always remained confident of our analytical skills. George and Wilma McInturff created the Southeast Center for Forest Economics Research (SCFER), the organization that drew many of us in the Research Triangle together to share and to develop critical insight into forest policy issues. Ralph Alig led SCFER in expanding its technical proficiency. He also provided particular assistance as an advisor and reviewer of our southern pine analysis.

Various others made important contributions along the way. Bob Eddleman and Doug Richards of the Cooperative State Research Service and Mississippi State University led a southern forestry advisory committee that oversaw the entire softwood plywood effort and encouraged our continued involvement in research evaluation. Hans Gregersen of the University of Minnesota and Dave Bengston of Minnesota, and later the US Forest Service, learned about research evaluation in parallel with us. They were our technical colleagues throughout. Vern Ruttan and Burt Sundquist of the University of Minnesota provided agricultural economics experience and kind advice.

Herb Fleischer and Jack Lutz, both retired from the Forest Products Laboratory, are largely responsible for the success of softwood plywood research in the 1960s and 1970s. They willingly shared historical and technical insights and reviewed our reorganization of softwood plywood research cost data. Gary Lindell found that data buried in a deep and dusty vault after we had given up all hope. Gary was a regular source of professional advice and cost data for both chapters 4 and 5. Various FPL researchers (Jeanne Danielson, David Lewis, David McKeever, Peter Ince, Jim Minor, and Lee Gjovik) assisted with background information on the sawmill, woodpulp and wood preservatives industries. Jeff Gill of Koppers Co. helped us to better understand the various wood preservative processes. Several Duke colleagues participated in preliminary analyses of the forest products industries: Tom Haxby (sawmills), Melinda Sallyards (woodpulp), and Allen Bruner and Jack Strauss (wood preservatives). Jim Baer provided data and computational assistance. Jim, like Gary Lindell, found a critical and otherwise unavailable data series.

Herb Knight, Ray Sheffield, and Rich Birdsey of the US Forest Service were thorough critics and kindly associates in our pursuit of southern pine growth and yield data. Steve Boyce, Duke; John Gray, Gifford Pinchot Institute; Bob Weir, Southern Forest Tree Cooperative and North Carolina State University; and Barry Malac, Union Camp; shared their observations and experiences in southern forestry research. Dwight Hair of the US Forest Service shared his historical cost series for Forest Service research. Richard Greenhalgh and Greg Alward, US Forest Service, gave freely of their extensive experience with the IMPLAN model. Tony Scott of the University of British Columbia sharpened our thinking about technical change in forest growth and management.

Ross Whaley of the US Forest Service, and later Syracuse University, encouraged us to consider the implications of our analyses beyond the confines of commercial forestry.[1] Peter Pearse, University of British Columbia, and Bill Bentley, Winrock International, encouraged us to consider these implications beyond the confines of the United States. Jim Douglas, World Bank, added a critical developing country insight. Roger

---

1. We restrict our discussion to wood and wood products in this volume. Nevertheless, Ross' point is an important one. Our comments on potential research productivity for the multiple of other forest land uses are contained in a chapter on demand-side management in C. Binkley, G. Brewer, and V. Sample, eds., *Redirecting the RPA,* Yale University School of Forestry and Environmental Studies Bulletin 95, 1988.

Sedjo, Resources for the Future, was once more the most thorough and reliable of critics. Dan McKenney and Jeff Davis, Australian Centre for International Agricultural Research, and David Brooks, US Forest Service, read our final chapter carefully and made improvements in it.

Barb Daniels was our invaluable research associate throughout. Barb was the core of the resource economics program at Duke during a time that was exciting for all of us.

Dolly Tiongco, Barbara Newman, and Wendy and Mike Seldon provided their own special enthusiasm, support and encouragement. Each has been a warm friend to all three of us. We owe them our most heartfelt thanks.

Finally, so that there is no mistake, we wish to identify only alphabetical rank among three equal co-authors.

WILLIAM F. HYDE
DAVID H. NEWMAN
BARRY J. SELDON

The Economic Benefits
of
Forestry Research

Introduction

THE EFFICIENT ALLOCATION of research dollars is a new topic of inquiry for forest and resource economists. Decisions having to do with budget allocations are usually made by politicians, managers and administrators. In recent years, however, tighter budgets for both public agencies and private corporations, greater scrutiny of the public planning process and better analytical tools have led to a more organized and disciplined inquiry into research budgeting. This raises natural questions for economics, itself a discipline of resource allocation. They are important questions because even small allocations to what becomes successful research can substantially reduce the effects of resource scarcity and improve social welfare. They are difficult questions, however, because forestry research is an activity on which prices work imperfectly, if at all. Fortunately economists from related fields, notably agriculture and industrial organization, have struggled with similar research budgeting and allocation problems and can assist us with their insights.

The profit motive has always justified internal scrutiny of *private* research budgets, but such scrutiny is more important during financially difficult times—as the years since 1978 have been for the forest products industries. The higher profile of *public* research budgeting can be traced from a variety of recent events. Critical issues in public agricultural research were the topic of a high-level workshop jointly sponsored in 1982 by the Rockefeller Foundation and the federal Office of Science and Technology.[1] The Renewable Resources Planning Act of 1974, as amended in 1976, and the Reagan Administration stimulated inquiry into public forestry research expenditures in particular. The former requires a decennial assessment of forestry in the United States, together with a planning document that includes planning for forestry research. The latter, through both the Assistant Secretary of Agriculture for Conservation and the Office of Management and Budget, expressed doubt regarding the benefits originating from public forestry research expenditures—thereby encouraging more careful measurement of these benefits. In partial response, the US Forest

Service established a major research work unit to explicitly inquire into this issue.[2] Simultaneously, two sections of the International Union of Forest Research Organizations identified research evaluation as their fundamental agenda item.[3] Finally, the Society of American Foresters acknowledged the timeliness of the topic in the theme, "Increasing Forest Productivity," of its 1981 national convention. Recognizing that research is the key to increasing productivity over time, that convention was the first to include sessions on research evaluation and research productivity.

This volume is a consequence of this background of interest. Our objective is to assess the economic returns and distributive impacts of public investments in forestry research. Economic returns may be non-monetary as well as monetary. Distributive impacts include both the allocation of research benefits and costs among various productive factors and the allocation between producers and consumers. Our focus on public research investments is a function of our public policy interest—and not a denial of the substantial research contribution of industrial forestry.

This introductory chapter first outlines the concepts involved in assessing economic returns and distributive impacts due to research and then reviews the potential justifications for a public research presence. The conceptual outline is background and preface for the detailed technical material in subsequent chapters. It will be repetitive for those familiar with the research evaluation literature but it may provide a necessary summary for foresters and for some policy analysts who find this material new. The justifications for forestry research are also most clearly understood with reference to the conceptual outline. A discussion of these justifications provides an important foundation for our empirical focus on public research investments in the subsequent chapters. Technical change is the output of research investment. A brief second chapter reviews historical measures of technical change, both in the forestry sector and in the US economy as a whole.

The remaining chapters are a more quantitative and empirical response to our fundamental objective. We investigate the gains from forestry research in various specific forest industries. Research examples from forest products (*e.g.*, wood utilization in the softwood plywood industry) and from timber growth and management (*e.g.*, the southern pine industry) serve as illustrative cases that provide a more general understanding of previous forestry research experience. These examples are also suggestive for future research budgeting and program planning. We will find that some forest products research has had a very high payoff. The returns on public research expenditures can range above 100 percent annually and the return on the marginal public research dollar invested exceeds $12.00 in particular industries. It may exceed $50.00. Our methods are consistent with a large body of agricultural literature and our *forest products* research results are generally consistent with agricultural results which suggest that, as a society, we have underinvested in research over the years. [See Ruttan (1982) for a survey.]

The economic history of *timber growth and management* research is less notable. Our chapter 2 review of technical change reflects that historic factor scarcity should cause us to anticipate greater incentives for, therefore greater gains to, forest products research rather than to timber growth and management research. Our illustrative empirical cases will support this expectation. Nevertheless, the final chapter reflects that the past importance of specific and selective forest products research projects may not be indicative of the future research payoff in the same industry. Indeed, the increasing relative scarcity of the *in situ* timber resource suggests a potential for increasing payoffs to future research in timber growth and management. Therefore, an important conclusion of this book has to do with understanding the significance of accurate *ex ante* research evaluations and the potential gains to society from allocating forestry research budgets according to an accurate interpretation of these *ex ante* instructions.

When society underinvests in research and the last dollar invested produces truly large gains (as in some forest products research) or virtually no gain at all (as in the historic case for timber growth and management research), then the incremental investment decision can be very important indeed. Thoughtful application of *ex post* research evaluation techniques such as ours can provide key assistance in this decision.

BACKGROUND CONCEPTS

There are two generally accepted methods for evaluating returns to research: the consumers' and producers' surplus/benefit-cost method and the production function method. We examine both methods briefly here, looking first at the consumers' and producers' surplus method with reference to the standard supply and demand functions as drawn in figure 1.1. Our empirical analyses of forest products research use a combination of the two methods and our analysis of southern timber growth and management research uses a variation of the production function method.

Figure 1.1 shows value per unit on the vertical axis and units of output on the horizontal. The quantity of product which consumers demand varies inversely with the product price according to the function D. The quantity that producers are willing to supply increases with their unit production costs according to the function S. The market for the product clears where the supply and demand functions intersect at point a, or where the price

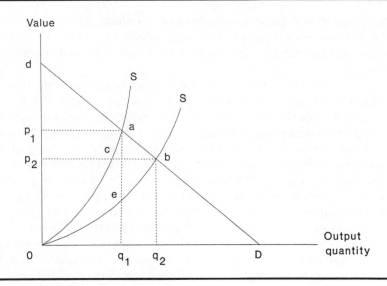

Figure 1.1. Consumer/producer surplus approach for analyzing
gains from technological change

consumers are willing to offer for the last unit of product, p_1, is also the
cost of producing that unit. This price induces a total of q_1 units of product
to exchange in this market.

The total production cost for q_1 units is the area under the supply
function up to the market clearing output level, or Oaq_1. This cost pays for
the variable inputs in the production process; raw materials, labor and
operating capital. All units of output sell at the same market clearing price
p_1. Therefore, all but the last unit attract payments greater than their
production costs. The total of these surplus payments, Oap_1, is the return
to the fixed factor of production and is known as economic rent or
producers' surplus.

Although consumers only pay p_1 for each unit of product they consume,
units consumed prior to the final unit are worth more to them. Total value
to consumers for q_1 units is the area under the demand function and to the
left of the market equilibrating output level, or Oq_1ad. Total value to
consumers, net of consumers' market payments, is known as consumers'
surplus p_1ad. The gain to society from producing and consuming q_1 units of
output is Oad, or the sum of the producers' plus consumers' surpluses.

The previous two paragraphs describe the static economic condition.
Technical change alters this situation. Technical change ("process"-oriented
technical change, to be precise) either alters the production process itself,
thereby permitting use of a lower cost combination of the same inputs, or

makes possible the use of less expensive substitute inputs. In either case, it decreases costs per unit of product such that the supply function shifts downward to, say, S' in figure 1.1. The new market equilibrium occurs at a lower price p_2 and a greater level of output q_2 clears the market. Technical change creates a gain in consumers' surplus equal to p_1abp_2 and a loss in producers' surplus equal to p_1acp_2 but a countering gain in producers' surplus equal to Ocb.[4] Technical change has no significant impact on the aggregate use of the inputs of this production process if the inputs all exchange in competitive markets and if the production process requires only a small share of the aggregate market availability of each input. In this case, those factors released from production by technical change find employment in other markets and at their unchanged market prices.[5] In conclusion, the net social gain due to technical change is $p_1abp_2 - p_1acp_2 + Ocb = Oab$. This approach to measuring the gains associated with technical change is also known as the "index number" approach because much of the literature concentrates on finding indices for measuring the relevant triangles in figure 1.1.

The consumers' and producers' surplus method permits benefit-cost calculations for research investments. The annual benefits are the net social gains due to technical change (described above and in figure 1.1) while the costs are those associated with doing the research, distributing knowledge of it, and implementing the technical change resulting from this research. The distribution, or diffusion, of the new technology over time and space creates major complexities for these benefit-cost calculations. That is, research and implementation costs occur in the initial investment year while benefits lag, occurring only with the gradual acceptance and implementation of the new technology. They also continue over time.[6]

Knowledge of the time streams of benefits and costs and the opportunity cost of capital together permit the calculation of a benefit-cost ratio for the particular research-induced technical change in question. Alternately, in the absence of a known opportunity cost of capital, research evaluations often search for the rate of return that sets the benefit stream equal to the cost stream. The result is an *average* rate of return for all dollars spent on the given research investment.

The second method for evaluating returns to research focuses on the production function. The production function explains the physical input-output relationship described by function F in figure 1.2. As the physical quantity of inputs increases, then physical output must also increase but eventually output per unit of input can increase only at a decreasing rate.

Technical change makes output increases possible from the same or fewer variable inputs as were used before introduction of the new technology. Figure 1.2 describes technical change as a shift of the entire production

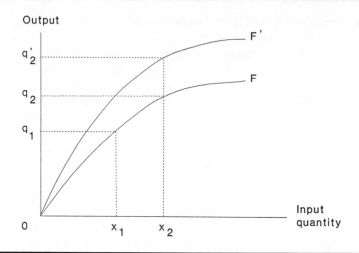

Figure 1.2. Production function approach for analyzing gains from technological change

function from F to F'. Observed output shifts reflect points on these functions. For example, the initial output level q_1 using x_1 units of input may shift to output level q_2 using x_2 units of input during a period in which technical change occurs (and quantity demanded increases).

It is important to distinguish a shift in the production function from a movement along the production function, a movement which explains expansion in output due to an increase in inputs and not to technical change. In the example above, the output adjustment due to a shift in the production function is $q_2' - q_2$ while the output adjustment due to a movement along the function is $q_2 - q_1$. Other data limitations force us to overlook this distinction between a shift in and a movement along a production function in our empirical analysis of timber production research in chapter 6. Therefore, the results in that analysis will be erroneous by the unknown equivalent of $q_2 - q_1$.

The empirical distinction between a shift in the function and a movement along the function is generally difficult to make over time because input quality often changes, thereby contributing to, and confusing measurement of, any gross change in the production relationship. Fully accurate measures of technical change over time require inputs adjusted to reflect constant quality per unit.

The production function method for evaluating research has the advantage of permitting calculation of *marginal* research impacts. For example, consider the generalized form

$$F' = F'(R,K,L) \tag{1.1}$$

where R, K, and L are physical units of research, capital and labor inputs, respectively. The marginal physical product of research is the first derivative of F' with respect to changes in its physical research input, symbolically $\partial F'/\partial R$. The value of the last research dollar invested, or the marginal value product of research MVP_R, is simply the marginal physical product times the price of a unit of output p_Q.

$$MVP_R = p_Q[\partial F'/\partial R] \tag{1.2}$$

If the production function represents production in one year, then MVP_R is also an annual measure. Nevertheless, once the benefits of research occur, they continue occurring year after year. They also form the basis from which future technical change occurs—and must be measured. Therefore, the marginal rate of return to research and development investments is that rate r which equates the discounted value of the perpetual stream of marginal research benefits with the cost of a unit of the research related input p_R.

$$\sum_{t=0}^{T} MVP_{R(t)} (1 + r)^{-t} = p_R \tag{1.3}$$

where $t = 0$ is the year in which the research investment occurs.

The Residual Approach

The first empirical problem has to do with the measurement of technical change, the output of research. Capital and labor must be measured at some initial time, say t_0. Research must also be measured from the initial year, but observed technical change occurs between the initial research year t_0 and a later given year t_1. The most early and elementary empirical evaluations of research, whether using the consumers' and producers' surplus method or the production function method, calculate technical change as a residual. They begin by estimating the functional relationship between output and known inputs at initial time t_0.[7] For the production function approach, this is

$$F(t_0) = F[K(t_0), L(t_0)]. \tag{1.4}$$

The second step in this approach estimates the production function at a later moment in time t_1 using the same functional specification and the new input levels known to obtain in t_1, $K(t_1)$ and $L(t_1)$, adjusted where necessary such that K and L refer to constant quality inputs.

$$F(t_1) = F[K(t_1), L(t_1)]. \tag{1.5}$$

Referring again to figure 1.2, this is the same as estimating the functional form F explaining the relationship between x_1 and q_1, and using this specification to determine the output level q_2' which would occur with the new input level x_2. The residual approach assumes that technical change shows up in neither estimate (1.4) nor (1.5) but explains the full difference between the estimated $F(t_1)$ from equation (1.5) and the observed output level at time t_1. The observed level is one point (q_2', x_2) on F' in figure 1.2.

Methodological Problems, Resolutions, and Extensions

These techniques for measuring research benefits raise several questions. The most serious has to do with attributing the entire unexplained residual to research and technical change. Improvements in the residual approach still leave unanswered questions about (1) changing input quality, (2) multiple inputs, (3) adjustments for lags between research inputs and their impact on productivity, and (4) research cost accounting. We might also inquire into the relationship between research evaluation and its recently popular cousin, technology assessment. The next pages respond to these questions in turn, beginning with a simple analytical improvement that permits direct estimation of research benefits to replace the residual approach.

The obvious objection to the residual approach to estimating technical change is that the difference between $F'(t_1)$ and $F(t_1)$ includes both technical change and the net effect of errors in the estimated contributions of other, non-research-related, productive inputs (including errors due to mismeasures of changing quality). The sum of these errors may be either positive or negative in sign and this sum biases any estimate of technical change accordingly. Our southern pine growth and management analysis will suffer from this problem.

The general solution to this problem is to regress observed output levels in the sample period on (a) observed constant quality labor and capital inputs, (b) *research-related variables* and (c) any terms explaining interaction between research and other inputs. The resulting regression coefficients for the research-related independent variables *directly* measure the impact of technical change embodied within the specific research input. So long as there is no statistical collinearity between independent variables, then there is no longer any risk of collecting net estimation errors, attributing them to technical change, and identifying them as payoffs to research. The more recent research evaluation literature, and our analyses of research in the forest products industries, make this correction.

Input quality: Increases in input quality may cause expanding output levels even in the absence of technical change in the production process.[8] Therefore, overlooking increases in input quality creates an upward bias in estimates of technical change or research impacts. Yet the observed norm is for labor and processed capital inputs to do just that, to increase in quality over time as the labor force becomes better educated and as machines become more efficient. On the other hand, unprocessed capital inputs or natural resources may decrease in quality over time as society turns to lower grade and less accessible sources. Thus, the decreasing quality of unprocessed capital and the increasing quality of labor impose opposite biases on measured technical change. In sum, input quality is something to watch closely in the empirical studies which compose the body of this book.

Multiple research inputs: As empirical analyses of research impacts add sophistication, they break the research term in their regressions ever more precisely into research itself and several related inputs. They may recognize both research and development, with development being the act of applying knowledge gained in research (Scherer 1980, p. 410). They may further divide research itself into basic and applied (Scherer 1980, p. 410) and into public and private research (Cline 1975, Ruttan 1982) depending on the requirements of the particular empirical study. They may also recognize that education occurs as the handmaiden of research and hastens the pace of its implementation (Welch 1971) and that, for agriculture, extension activities play a large role in development. Finally, they may recognize that, over time, new political regulations on production can interfere (Oster and Quigley 1977), constraining what otherwise could be substantial technical change and thereby constraining payoffs to research-related inputs. Not all of these distinctions in research-related inputs or constraints on their impacts are included as independent variables in each empirical analysis, nor need they be. Many would be unimportant for any given product. It is important, however, to recognize these distinctions and to include the various research-related inputs where each is appropriate to the production process and to the objective of the particular analysis.

Questions of which research-related inputs to include in the regressions assessing technical change also raise similar questions regarding cost accounting. The full cost of technical change includes the entire sequence of costs, from those associated with basic research through those associated with applied research to development costs. In application this often means public expenditure for basic research, private expenditure for modifications to satisfy output objectives and private expenditures for development, where development suggests modifying or exchanging existing equipment and retraining labor.

The general experience of the US Forest Service's Forest Products Laboratory (FPL) provides a useful example. The FPL generally undertakes risky and longer-term research projects having to do with new product development or potential improvements in mill operations. Firms from the forest products industries generally begin to show interest only after the FPL produces initial evidence of research success. By then, both the time period between initial research and final application and the uncertainty of research success have decreased. Subsequently, research units within industrial firms take the public information gained from the FPL's initial research experience and begin a race with each other and the FPL to develop the applied patents. Finally, the firms make the necessary modifications in their mills to implement the newly patented production process or to produce the newly patented product.

In other cases more common to the experience of agriculture and more relevant to timber production by non-industrial private landowners, the public sector is the source of funds for both basic and applied research. The public sector is also the source of funds for forestry extension activities—with their educational emphasis. Of course, there may be varying federal and state shares in research and development (R&D)—with the states often becoming relatively more involved in extension. Extension agents inform the non-industrial landowners of the new production information and landowners then modify this newly acquired knowledge to fit their own timber growing objectives and their own existing forest operations.

Research and implementation lags: The sequence of the benefits and costs of technical change raises questions of the time lags between initial basic research and final implementation and of the diffusion of successful technical change over time and space. These are empirical questions and their answers vary from case to case. They are important because a few years' delay in implementation can reduce substantially the discounted benefits of a given technical change.

We might anticipate that lags between research and implementation are shorter in concentrated industries where a few single firms have large stakes in the gains from a new technology. Therefore, we might anticipate shorter lags in the relatively more concentrated pulp and paper industries than in the lumber industry—which is composed of many small sawmills (and a few larger ones). We also might anticipate shorter lags in the forest products industries in general than in the timber management industry, especially where much of the timber management occurs on small ownerships of non-industrial forestland. Smaller owners are usually more isolated than larger owners from the most recent information about research results and technical change. Indeed, the research-implementation lag in the forest products industries is virtually equivalent to the research-productivity lag because production runs are generally short. Several usually occur in each day. This contrasts with the timber management industry where implementation implies on-the-ground application of new technologies, but implementation may precede its effect on eventual harvests by several decades. Our empirical analyses in chapters 5 and 8 support these hypotheses.

Cost accounting: The cost of technical change is the total cost of all of these R&D related operations combined. Our empirical examinations of federal research must recognize that federal expenditures are only one component of the total costs of technical change. Therefore, the costs of this component should be compared with only their appropriate share of total technical change, a share which can be found using proper regression techniques and two research-related independent variables, one for federal research expenditures and the other for all other R&D. This technique avoids the temptation to count all of a given technical change as the benefit resulting from a single component of all costs, thereby unjustly ignoring, in this example, the efficiency of state and private research-related expenditures and inflating the efficiency of federal expenditures.

It is not unreasonable, however, to compare each independent research or development cost with its own share of the benefits of the given technical change. Doing so only requires care in not overlooking any R&D inputs and in avoiding double counting of any of the benefits of technical change.

Technology assessment: Technology assessment is not a major objective in this book. Nevertheless, technology assessments are a closely related idea and producing technology assessments has become an interesting activity for researchers and consultants in recent years. The Congressional Office of Technology Assessment has been particularly active, but US Department of Agriculture agencies also request technology assessments from time to time. The purpose of a technology assessment is to show the impacts of a given production technology on energy, employment, the environment or whatever social, economic and political decision criteria may be important to local communities affected by the location and use of the technology in question.

The relationship between technology assessment and research evaluation is easily understood with reference to figures 1.1 and 1.2. Technology assessments measure the impact of a given technological change—often an anticipated change, not an actual one—shown by a shift in the supply function from S to S' and a shift in market equilibrium from a to b in figure 1.1 or a shift in the production function from F to F'' in figure 1.2. Technology assessments require a measure of the gain displayed by these shifts. Therefore, they are related to the benefit measurements in research evaluations. When technology assessments ask about energy, employment, the environment or whatever, they are asking about changing use levels for these factors of production due to the specific technology in question. Supply and production functions measure the impacts of the aggregate of all productive factors. Therefore, answering these questions is conceptually simple. We need only to examine the component parts of the measure of technical change or of its underlying supply and production shifts or, equivalently, the capital and labor inputs in equations (1.1) through (1.5).

Of course, the empirical difficulties are that these inputs may have to be more precisely specified than just "capital" or "labor" and the specification may be particularly troublesome if (1) the distribution of a given input among affected populations is important or (2) some inputs are not readily measurable or, even if measurable, are open access resources of uncertain value—such as environmental inputs like air and water. Chapter 7, which reflects on the distribution among factors of gains to southern pine growth and management research, is one example of a technology assessment.

A ROLE FOR PUBLIC RESEARCH INVESTMENTS

In a volume devoted to assessing the benefits of public research, it is appropriate that we also review the underlying justifications for a public role in research.

There is much general comment about the role of public agencies in forestry research. There is even considerable discussion about whether there should be a role: Perhaps the gains to forestry research are all captured in the market by private firms capable of conducting their own research. To our minds, there is little clear thought devoted to this topic. Therefore, it may be useful to outline the rationale for this kind of public market intervention.

Three valid economic arguments or cases can justify public research effort. First, private investors may be unwilling to invest in research projects where the expected research output is uncertain, where there is a long time horizon between the initial expenditure and the anticipated final payoff from implementing the research breakthrough, or where the necessary initial research investment is truly great. This may be true even where the research investment is efficient in the general sense that it has a positive present net worth, benefit/cost ratio greater than one, or rate of return greater than the firm's guiding rate of return. Tree improvement is an example of a forestry research investment with a long delay before the eventual payoff, yet a research investment that may be efficient. (There probably are no forestry examples where the initial research investment is too great for private involvement. This argument is probably best used as a justification for large and uncertain public research investments in space technology or for improving national defense.)

Reference to supply functions S and S' and demand function D in figure 1.1 aids the discussion of the remaining two cases. Both are important in forestry. Research breakthroughs cause a decrease in production costs—or a downward shift in the supply function from S to S'. The market clearing price-quantity relationship shifts from p_1,q_1 to the lower price and greater quantity shown by p_2,q_2. Consumers gain p_1abp_2. Producers lose p_1acp_2 but also gain Ocb which together may or may not be sufficient to make a positive aggregate impact on producers. In all cases, the aggregate social gain (the sum of consumers' and net producers' gains) Ocb is positive. Since gains to private investors (producers) by themselves may not be positive, then private investors may have no incentive to invest and a public presence may be both necessary and justifiable on basis of the positive net social gain.

In one case, the demand function is relatively inelastic (close to vertical) and the supply function is relatively elastic (close to horizontal). Virtually all benefits accrue to consumers; producers may even be net losers. In this case, private industry is unlikely to conduct research. The sawmill industry is characterized by demand and supply functions similar to these and, not surprisingly, we observe little private research conducted by this industry. Consumers gain substantially, however, from the lumber price decreases created by sawmill research and technical change.

In the final case, producer gains are positive but there are many firms in the industry. Dividing the aggregate gains among each firm does not leave a sufficiently large gain to pay for the research investment if each firm has to conduct its own research and, thereby, duplicate each other firm's investment. It is not surprising, therefore, that industries with large research budgets are also industries with either few firms among which to divide the research benefits or industries with clear product identification, therefore with enforceable proprietary claims to research breakthroughs. The drug and chemical industries are examples of the latter.

The lumber, furniture and timber management industries may all suffer from the problem of sharing research benefits and duplicating research costs. We will learn, in chapter 4, that the softwood plywood industry may suffer similarly. For the southern forest industry, the various industry/university cooperatives may be examples of one private market remedy to the problem. These cooperatives pool research expenditures and share research results, thereby avoiding duplication.

The patent system attempts to protect a return on private research investments in a further example for this latter case. Patents intend to restrict the research gains to the initial successful innovator as a property right, but they do not always succeed. Rapid duplication around the patent may dissipate all patent holder and private producer gains—thereby justifying a public role in research.

That is, a successful patent holder can sell rights to the patent to one firm for a value as great as area *Oae* in figure 1.1. This firm produces less expensively (at costs reflected by S') but it must also recover its fixed patent purchase costs. The market will remain near the old equilibrium p_1, q_1. If, however, other firms can quickly copy the innovator by designing something similar, then the patent becomes worthless to the patent holder. The market shifts to a new equilibrium at p_2, q_2 and all producers together share the gain equal to area *Ocb*. The portion of that gain equal to area *abc* immediately passes on to consumers—to be added to the consumers' gain of $p_1 acp_2$. The innovator with the original patent receives nothing. The producer or patent developer receives gains only when one producer can maintain a patent advantage over other producers for long enough to recover the private costs of the innovation. If the producer has no incentive to innovate, then society loses the benefits of innovation unless the public sponsors the research.

The powered back-up roll in plywood mills could be an example of this final case. It is a simple but effective idea, developed at the publicly-funded Forest Products Laboratory, which places uniform pressure on logs being peeled for plywood—and increases yield per log by as much as 17 percent. If the powered back-up roll had been a privately funded development, then we might anticipate that once this idea became known many firms would have developed the technology for themselves and there would have been minimal return for the private innovator. Nevertheless, considerable gain did spread throughout the industry as a whole and then passed on to consumers, where it persists.

A final caveat is in order before we consider our actual measurements of research gains in forestry: These preceding three arguments may provide justification for public activity in a specific research area. They are not automatic justifications for all public research. Each case of public research must be examined on its own merit in terms of these three justifications. If the specific research investment appears to have merit, then it must still pass the test of social efficiency. That is, the research investment in question must still obtain a return which, at the margin, equals or exceeds that anticipated for other investments, research or otherwise, public or private.

ORGANIZATION OF THE BOOK

The empirical analyses of chapters 3 through 8 provide evidence for the various hypotheses raised in this introductory chapter and in the historical review in chapter 2. The main theme of the empirical chapters is the effectiveness of forestry research expenditures. We will calculate the economic return to dollars spent in research. We will also comment on the distributive impacts of research expenditures where the data and analytical method permit—focusing on distributive impacts among factors, landowner or other producer classes, geographic regions and consumers.

Our general objectives are twofold: to demonstrate the methodology as it necessarily changes for various analytical cases and to derive policy-oriented conclusions. Our analytical methods call on the experiences of agricultural economics modified to fit the requirements of available forestry data.

These forestry data are often less precise as to geographic origin and as to specific factors of production than are comparable agricultural data. Furthermore, they may have been collected for a shorter total period of time. This latter difference is of particular importance when one considers the large differences in length of the production periods between forestry and agriculture; therefore, the general requirement for longer sequences of forestry data in order to draw conclusions equally as valid as those drawn from reliable agricultural data. The timber management industry has the additional problem of overlapping production periods but data which do not explain the overlap. That is, trees of many age classes, therefore of many planting and expected harvest years and many production technologies, are indistinguishable in basic forest inventory data. Yet comparable production data from agriculture and other industries reflect the production runs of only one year, one season or even part of one day. Our intention is to make the methodological modifications required by these data differences of a general nature they may serve as a guide for our empirical analyses as well as for subsequent evaluations of other timber management research projects.

The convenient methodological and policy divisions are between industry specific (*e.g.*, sawmills, plywood and veneer mills, etc.) studies and aggregate or forestry-wide studies where, in both cases, analyses of research impacts are of either an *ex post* or an *ex ante* nature.

Ex post analyses provide critical review of previous investments in forestry research and should be conducted with the expectation of learning from previous experience. They show broad categories of historic success and failure. They, together with careful reflection, may suggest the characteristics of successful future investments. *Ex ante* analyses explicitly anticipate the benefits and costs of forthcoming research investment alternatives. They are conducted with an eye on the planning process.

The original intent of chapters 3-5 was to examine the impacts of previous (*ex post*) narrowly defined and explicit research expenditures. It is difficult, however, to separate the impacts of explicit research projects when research may proceed simultaneously on several distinct projects within a given industry. Neither research inputs nor research outputs associated with a single innovation can be identified easily.

A reasonable alternative is to assess the collection of all research conducted in a particular industry, but to define the industry conducting research as narrowly as data permit, thereby restricting the likelihood of measuring benefits to multiple research projects. Therefore, in these three chapters, we examine the softwood plywood industry and three 4-digit SIC code industries. Our method of analysis extends the previous research evaluation literature (1) by using the dual of the supply function as a means of combining the consumers' and producers' surplus method with the production function method and (2) by permitting all factor costs to vary throughout the period of inquiry. The softwood plywood chapters explain the method in detail. These chapters can serve as a useful model for both the remainder of this book and for future work by others.

Chapter 5 more briefly reviews research impacts in the three 4-digit industries: sawmills, woodpulp and wood preserving. This chapter concentrates on analytical results, rather than on methods. Chapter 5 also takes these results one step further and reflects on the Schumpeterian question about research and firm size: Do firms of a larger or smaller absolute size expend greater research effort? Or does firm size have little to do with research expenditure while market share may be a better predictive variable? (Schumpeter 1950, ch. 2; Scherer 1980, ch. 15; Kamien and Schwartz 1982, pp. 27-29). Observations from this chapter may provide further reflection on which forest product industries are most likely to conduct their own (private) research and which may benefit most from greater public support and encouragement.

Chapters 6 through 8 examine southern pine research. Their contributions are their timber growth and management research focus and their aggregate nature. These are the chapters that pose the greatest methodological difficulties. Infrequent data collection and overlapping production periods for timber management create analytical problems with which agricultural economists have not had to deal and econometricians would prefer to avoid.[9]

Chapters 3 through 5 discuss public research in specific forest products industries. Chapters 6 through 8 examine the aggregate impact of a wide range of research (*e.g.*, fertilization, tree improvement, forest management, growth and yield, etc.) for a broad segment of the timber management sector, the dynamic southern pine industry. This aggregate analysis has the merit of neither overemphasizing the success nor disregarding the failures of single research projects or even broad research categories.

Southern pine research began early in the twentieth century and includes fire control research expenditures made since the 1930s. Productivity data are weak this far back, however. Therefore, the analysis in chapters 6 through 8 proceeds only with some skepticism. Nevertheless, chapter 8 shows that even generous technical change and research productivity estimates and conservative research cost estimates suggest historically small net southern pine research yields. The southern pine species group and its region of the country are generally thought to provide the best example of successful timber growth and management research. Therefore, more general, US-wide returns to timber management research have probably been smaller yet.

These are all *ex post* analyses. They can tell us what has been and perhaps they can throw light on the characteristics of previous research successes and failures. They do not directly and immediately predict the future of forestry research by themselves. The chapter 2 survey of our expectations for technical change in forestry and some of our policy oriented summary comments in the concluding chapter provide our attempts to be forward looking. We speculate that biological research investments in US forestry may become more important as the American forest economy matures in the early part of the twenty-first century.

There is little use for an aggregate and *ex ante* analysis because aggregate planning decisions generally are based on either past reputation—as tested here in the aggregate and *ex post* timber management case in chapters 6 through 8—or the sum of expected good decisions on proposed specific current and future research projects. The *ex ante* comments on timber management research investments in chapter 9 reflect the need for the latter.

Our path through the various basic analytical and policy categories in forestry research fits the following format:

	ex post	*ex ante*
Industry specific		
Forest products	Chs. 3-5	Bengston (1984)
Timber management	N/A	Chs. 2, 9
Aggregate		
Forest products	—	N/A
Timber management	Chs. 6-8	N/A

We argued that there is no policy usefulness for aggregate *ex ante* studies. Our chapter 8 observation of low returns for previous southern pine research will suggest that there is also little point to analyzing the specific sub-categories of southern pine research. If these produced very low returns in aggregate, then most of them produced unspectacular returns individually. Therefore, there is little reason to examine most *ex post* specific-timber management research categories. Bengston (1984) examines one of the five remaining categories in the table. We examine three more. No one has examined the *ex post* aggregate case for forest products. [A method similar to ours in chapters 3-5, but applied to the broad SIC codes 24-26 (for lumber and wood products, furniture and pulp and paper), is probably appropriate.]

The final chapter summarizes our findings and draws conclusions. It has policy implications for research planning and expenditure in both public and private forestry. In particular it calls on general insights from previous empirical chapters to suggest the characteristics of successful research projects, their impacts on local economies and the justifications for both public and private research activities. Who are the great benefactors of public forestry research? Is a continuing public presence justified—or is forest industrial research a private good whose benefits are generally captured by the innovators themselves?

One important general finding will be that forestry research science, particularly forest products research, has produced returns to public investments that are much greater than the returns anticipated from most direct timber management activities. Furthermore, public forest products research in its role as a substitute for various other factors of production has had a larger social impact than many issues with considerably higher political profiles. For example, several forest products research projects have each had greater positive impact on timber availability than the potential negative impact of all proposed wilderness withdrawals together![10]

These observations raise questions for further analysis. The obvious questions have to do with whether we should expect to observe these high *ex post* returns to persist across the aggregate of all forest products industries and whether we can expect similar high returns to persist in the future, and for which industries. The final chapter addresses these questions. It is important for resource allocation and social welfare that we follow the instructions of this chapter. The chapter closes with our justifications for anticipating increasing payoffs for future timber growth and management research, in the US and around the world.

NOTES

1. See "Science for Agriculture: Report of a Workshop on Critical Issues in American Agricultural Research" published by the Rockefeller Foundation, October 1982.

2. Research Work Unit FS-NC-4204, Methods for Evaluating Forestry Research, under the direction of C. Risbrudt when the research for this book began.

3. S6.06-01 on Evaluation of Forestry Research, initially under the coordination of R.Z. Callaham and S4.05-01 on Evaluation of the Contribution of Forestry to Economic Development, initially under the coordination of A.J. Grayson.

4. Technical change has a negative effect on producers' surplus ($Ocb < p_1acp_2$) when demand is relatively inelastic and supply is relatively elastic. In this case producers have no incentive to innovate, regardless of the magnitude of net social gains, and public investment may be the only financial source of research. This is an important special case for forestry because the demand and supply for lumber are generally thought to meet these elasticity conditions.

5. In the event that all released factors of production cannot find new employment, then we might examine the displacement of area Oae in figure 1.1. Schmitz and Seckler (1970) provide an example with their classic article on the increasing social importance of research leading to development of the tomato harvester. The released factor of production was migrant labor.

6. More precisely, the benefits of research (the implementation of and production from new technologies) spread (or "diffuse") as sigmoid functions of time and space. The diffusion literature focuses on rates of adoption and identification of critical points in the sigmoid diffusion functions. Rogers (1983) is a basic text in the field. Buongiorno and Oliviera (1977) is the only forestry article familiar to us.

7. Explanations of this process for the consumers' and producers' surplus and the production function approaches are comparable. The original production function/residual example is sufficient for our discussion.

8. Changes in input quality are also technical change, but they represent a different technical change than we are interested in measuring. For example, when measuring research that causes technical change in the sawmill industry we wish to control for changes in rural education which permit the sawmill industry to employ a higher quality work force. Similarly, we wish to control for technical change in the steel industry which creates longer lasting steel for saw blades.

9. There have been a few previous evaluations of timber production research. (Footnote 3 in chapter 2 and table 3.1 in chapter 3 provide lists.) None are

aggregate and most are speculative rather than empirical. The data problems discussed in this paragraph have proven difficult to overcome.

10. In one graphic example, conversion of only 50 percent of all 1982 new US residential housing to the truss frame construction technique saves a volume of wood equivalent to the annual programmed harvest on all commercial forest lands (28.6 million acres) proposed for wilderness withdrawal in RARE II. (Source: USDA Forest Service calculations provided by H. Gus Wahlgren, February 15, 1989)

CHAPTER 2

A Short History of Technical Change

TECHNICAL CHANGE is the generalized product of research. That is, research is the input and technical change is an intermediate product which, together with other inputs (land, labor, and capital), fuels the economy. Its impact is measured as the change in economic output which would occur if both the quantity and quality of all other inputs were held constant. (Recall figures 1.1 and 1.2). We begin this chapter with a few words about the general technical change literature before turning, first, to empirical findings in the returns to research literature and, then, to a closer look at technical change in forestry in particular.

TECHNICAL CHANGE AND THE US ECONOMY

Making the best use of resources at any moment in time is important but, in the long run, it is dynamic performance that counts. For example, consider an increase from 3 percent to 5 percent per annum in the rate of economic growth due to technical progress. This increase overcomes, in just five years, the effect of a static output constraint causing a 10 percent reduction in potential gross national product. Thus, we can say that factors contributing to economic *growth,* in time, explain a large share of the *static* level of aggregate output as well. Furthermore, in time, growth may replace the impacts of all public restrictions (whether environmental, national security, income distributive or political) on static economic activity.

Research is one factor contributing to economic growth, perhaps the most important factor. For example, the gains in timber utilization originating from developments such as structural particleboard or truss-frame housing may very quickly offset a permanent restriction on, say, herbicide usage in timber production or on harvests from designated wilderness lands. Publicly funded research at the Forest Products Laboratory was instrumental in the development of both structural particleboard and truss-framing.

The underlying significance of technical progress or technical change is that it substitutes new knowledge for less abundant and more expensive resources: land, labor, and capital. Research is the input which produces this new knowledge and, thereby, creates increased returns to land, labor, and capital either by decreasing their costs or increasing their outputs ("process" innovation) or by increasing the quality of production through the introduction of new production processes ("product" innovation).

Solow (1957) responded to the ideas of these last paragraphs. From 1909 to 1949 the measured average increase in nonfarm productivity per worker-hour was 1½ percent per annum in the United States. Solow finds that 19 percent of this 1½ percent was due to expanding capital intensity (*i.e.,* more machines per worker). The remaining 81 percent of annual growth in nonfarm productivity was due to (1) technical improvement in productive inputs and productive practices (technical change in its strictest sense) and (2) increased quality of the labor force. Denison (1974) extended Solow's analysis for the years 1929 to 1969. He finds that 22 percent of nonfarm productivity growth per worker-hour was due to improved education of the labor force, 48 percent to advances in scientific and technical knowledge and only 12 percent to expanding capital intensity. Solow's and Denison's conclusions are that growth in output per worker in the US predominantly originated from the application of ever newer, superior production techniques by an increasingly well-trained work force.

Furthermore, their conclusions are conservative to the extent that Solow and Denison only measure process innovation and not product innovation. There is no satisfactory measure of output quality changes. Yet quality changes, and therefore product innovation, clearly occur. They are a partial result of technical change and improved labor quality, both of which are products of research. Therefore, Solow's and Denison's impressive measures of the impacts of technical change are *underestimates* of the impacts of research-related activities on economic growth (Usher 1964).

Knutson and Tweeten (1979), extending the original methodologies of Evenson (1968) and Cline (1975), made a related inquiry into the productivity of research expenditures. They restrict their attention to agriculture, specifically to all production-oriented agricultural research and extension for the years 1939 to 1972. They find a 50 percent annual return on the marginal research and extension dollar spent in the decade from 1939 to 1948. This rate of return has tapered off since 1948. Nevertheless, it remained an impressive 35 percent for the years 1969 to 1972, the final period in the Knutson-Tweeten analysis.[1]

These three studies examined either economy-wide technical change or the impacts of aggregate research expenditures. There also have been numerous analyses of the impacts of more specific research activities, many of them agricultural research activities.[2] Griliches (1958) initiated this micro-analytic work with his investigation of the social returns to research expenditures on corn hybridization. He finds an annual internal rate of return in the neighborhood of 35-40 percent for the period 1940 to 1955. Brumm and Hemphill (1976) and Ruttan (1980) surveyed the subsequent literature. Table 2.1 reproduces the summary of Ruttan's more recent survey. Ruttan's categorization of the literature as either "index number" or "regression analysis" corresponds to the consumers' and producers' surplus and the production function methods of research analysis. Ruttan finds 36 articles reporting returns that range from 11 percent for wheat research in Colombia to 110 percent for both cotton research in Brazil and rapeseed research in Canada. The majority of the articles in Ruttan's survey report returns exceeding 35 percent per annum regardless of the research problem or the country that serves as focus for the research. The Brumm and Hemphill survey includes nine articles on non-agricultural research problems. The results in these nine articles are consistent with those observed in the Ruttan survey.

These rates compare most favorably with private industrial internal rates of return which typically average 10 percent or more on capital investments and fall to the range of 2 to 5 percent at the margin. Thus, in spite of the difference between social and private returns, possible measurement and estimation errors, and various problems inherent in comparing rates of return that sometimes may be more average than marginal, the magnitude of difference between rates of return for research, particularly agricultural research, and rates of return for general capital investments presents an impressive case for the general success of research investments. It is persuasive evidence of the underallocation of resources to research. It becomes all the more persuasive when we recall that these are returns for process innovation. They overlook the additional but unquantified benefits of research leading to new or improved products.

In sum, apparently there has been opportunity for considerable social gain from additional investments in research in general. The recent historical evidence is so strong that we safely anticipate similar comparative results for *current* research expenditures. We anticipate with particular conviction because current research budgets seem so constrained. Many more dollars could be spent on research before its marginal return would fall to a level comparable to that of marginal investments in other sectors of the economy.

The natural question for us in this book has to do with whether forestry research generates similarly high rates of return, thereby suggesting a similar underallocation of resources to forestry research and a similar social gain to expanding forestry research budgets. Previous attempts to respond to this question have been scarce.[3]

Table 2.1. Summary studies of agricultural research productivity

study	country	commodity	time period	annual internal rate of return (%)
Index Number:				
Griliches, 1958	USA	Hybrid Corn	1940-1955	35-40
Griliches, 1958	USA	Hybrid sorghum	1940-1957	20
Peterson, 1967	USA	Poultry	1915-1960	21-25
Evenson, 1969	South Africa	Surgarcane	1945-1962	40
Barletta, 1970	Mexico	Wheat	1943-1963	90
Barletta, 1970	Mexico	Maize	1943-1963	35
Ayer, 1970	Brazil	Cotton	1924-1967	77+
Schmitz and Seckler, 1970	USA	Tomato harvester, with no compensation to displaced workers	1958-1969	37-46
		Tomato harvester, with compensation of displaced workers for 50% of earnings loss		16-28
Ayer and Schuh, 1972	Brazil	Cotton	1924-1967	77-110
Hines, 1972	Peru	Maize	1954-1967	35-40 50-55
Hayami and Akino, 1977	Japan	Rice	1915-1950	25-27
Hayami and Akin, 1977	Japan	Rice	1930-1961	73-75
Hertford, Ardila, Rocha, and Trujillo, 1977	Colombia	Rice Soybeans Wheat Cotton	1957-1972 1960-1971 1953-1973 1953-1972	60-82 79-96 11-12 none
Pee, 1977	Malaysia	Rubber	1932-1972	24
Peterson and Fiszharris, 1977	USA	Aggregate	1937-1942 1947-1952 1957-1962 1957-1972	50 51 49 34
Wennergren and Whitaker, 1977	Bolivia	Sheep Wheat	1966-1975 1966-1975	44 -48
Pray, 1978	Punjab (British India) Punjab (Pakistan)	Agricultural research and extension Agricultural research and extension	1906-1956 1948-1963	34-44 23-37
Scobie and Posada, 1978	Bolivia	Rice	1957-1964	79-96
Pray, 1980	Bangladesh	Wheat and rice	1961-1977	30-35

Table 2.1. (*continued*)

study	country	commodity	time period	annual internal rate of return (%)
Regression Analysis:				
Tang, 1963	Japan	Agreegate	1880-1938	35
Griliches, 1964	USA	Aggregate	1949-1959	35-40
Latimer, 1964	USA	Aggregate	1949-1959	not significant
Peterson, 1967	USA	Poultry	1915-1960	21
Evenson, 1968	USA	Aggregate	1949-1959	47
Evenson, 1969	South Africa	Sugarcane	1945-1958	40
Barletta, 1970	Mexico	Crops	1943-1963	45-93
Duncan, 1972	Australia	Pasture Improvement	1948-1969	58-68
Evenson and Jha, 1973	India	Aggregate	1953-1971	40
Cline, 1975 (revised by Knutson and Tweeten, 1979)	USA	Aggregate	1939-1948	41-50
		Research and extension	1949-1958	39-47
			1959-1968	32-39
			1969-1972	28-35
Bredahl and Peterson, 1976	USA	Cash grains	1969	36
		Poultry	1969	37
		Dairy	1969	43
		Livestock	1969	47
Kahlon, Bal, Saxena, and Jha, 1977	India	Aggregate	1960-1961	63
Evenson and Flores, 1978	Asia-- national	Rice	1950-1965	32-39
			1966-1975	73-78
	Asia-- internt'l.	Rice	1966-1975	74-102
Flores, Evenson, and Hayami, 1978	Tropics	Rice	1966-1975	46-71
	Philippines	Rice	1966-1975	75
Nagy and Furtan, 1978	Canada	Rapeseed	1960-1975	95-110
Davis, 1979	USA	Aggregate	1949-1959	66-100
			1964-1974	37
Evenson, 1979	USA	Aggregate	1868-1926	65
	USA	Technology oriented	1927-1950	95
	USA	Science oriented	1927-1950	110
	USA	Science oriented	1948-1971	45
	Southern USA	Technology oriented	1948-1971	130
	Northern USA	Technology oriented	1948-1971	93
	Western USA	Technology oriented	1948-1971	95
	USA	Farm management research and agricultural extension	1948-1971	110

Source: Ruttan, 1982

A HISTORICAL PERSPECTIVE ON TECHNICAL
CHANGE IN AMERICAN FORESTRY

It may be reasonable to review the history of forestry in the United States before we begin our empirical estimation. Understanding this history may help us anticipate both the locus of successful forestry research in the past and some of its future characteristics as well.

Until the last 35 years, American forestry could be characterized by the frontier model of economic development. That is, land was relatively more plentiful than labor as a factor of timber production and transportation costs were not a substantial deterrent to harvesting. Expanding harvests rather than increasing growth accounted for most increases in output. As a result, the conservation model of forest development, emphasizing notions of crop husbandry and soil depletion, never fully transferred from Europe to the United States. High and rapidly rising wage rates characterized the American economic environment and advances in mechanical technologies, particularly for timber harvest and utilization, dominated more labor-intensive conservation-oriented timber management practices.

This describes our forest economy (only fire control may have been an exception) since the Jamestown settlement. American forestry is only now beginning to change from a frontier economy to a developed economy where land is also a scarce factor of production.

This history is most easily seen by following the geographic expansion of forestry across the country. North America presented its European colonists with an immense natural endowment of mature timber. Harvests of this timber concentrated in New England from the time of colonization through the mid-nineteenth century. The great Lake States pinery was harvested in the late nineteenth century. The South harvested substantial volumes throughout the nineteenth century and it became the leading timber producing region in the first third of the twentieth century. Its production remains strong today. Southern forests have always had the advantages of easy access (to markets and to the timber itself) and rapid timber growth. West Coast harvests also became important in the early twentieth century. Here, the trees were large and easy water-borne transportation was available. Western harvesting gradually moved down the coast from Seattle and up the coast from San Francisco. The Rocky Mountain and Intermountain West were skipped over because plentiful timber remained in other, more populous regions. Moreover, Rocky Mountain and Intermountain timber stocking was less dense, access was more difficult, and transportation costs to the large eastern markets were great. Harvests in these latter regions are still not large relative to some other regions, although they have become somewhat more important since

World War II.[4]

These geographic trends are a generalization, of course. There has always been some harvesting for local use within each region. Even within regions, however, harvests have generally followed the frontier model; from more to less accessible land, from better to poorer quality species, from better to poorer stocked timberstands, and from higher to lower grade trees. Many fringe trees are left as Nature's bounty at the margins of productive land in each region even today—although nowhere is their physical inventory as great as it once was.

Thus, timber production has been characterized by *harvesting* throughout the 375 years of European development of North America. There has been little incentive to expend scarce resources *growing* timber while a bounty of mature timber remains. The mature timber, however, is nearly gone and the period characterized by concentration on harvest technologies may be drawing to a close. This event is, perhaps, hastened by public agency policy and recent environmental restrictions on timber harvesting and land use, restrictions which generally make it more difficult to harvest the remaining mature timber on public lands.[5] In other words, the timber frontier is closing, timberland is becoming relatively scarcer (more valuable), and the attention of timber producers may be turning to plantation management and more intensive forms of timber growing. Nowhere is this trend more apparent than in the South which is becoming the leading timber producing region in the world.

The long-term upward relative price trends of stumpage and sawlogs support this view (Ruttan and Callahan 1962, Barnett and Morse 1963, Phelps 1977, Manthy 1978). We might expect these trends to continue as long as a stock of mature timber remains. Indeed, we might expect the rate of relative stumpage price increase to reflect the rate of depletion of the stock of mature timber (Lyon 1981). Berck (1978), Libecap and Johnson (1978), and Sedjo and Lyon (1987), in independent analyses, report empirical results consistent with this expectation.

Rising relative stumpage prices are entirely consistent with the expectations of the economic theory of stock resources—which is best known for its application to hardrock mineral extraction. The standing mature timber available for the early colonists was like a newly exposed mineral deposit that comprises a large share of all of that particular mineral available at some moment in time. Availability of this new deposit sharply and immediately depresses the market price of the mineral. As extraction gradually depletes the new mine, only deeper and lower quality material remains. The mine's production costs increase. Eventually, the mine loses its competitive production cost advantage over other mines, and the market, including production from all other deposits of this mineral, stabilizes at some new natural price level.

Similarly, one expects stumpage prices to increase as long as there is a natural stock of mature timber and as long as the harvesting of this stock can proceed to less accessible sites, less densely stocked stands, poorer quality species, etc. The rate of price increase will eventually taper off and stumpage prices will find their natural level when most of the mature timber is gone and the costs of growing timber equal the value of timber produced. The small remaining volume of mature timber in this country and the financial successes of southern timber plantation management suggest that this will occur soon. Berck (1978), Adams and Haynes (1980), and Sedjo and Lyon (1990) project that it may occur within the next 30 to 50 years. (Southern pine prices may have already stabilized.) In other words, they expect constant relative stumpage prices within the next timber rotation.

The important questions for us are what this history suggests for technical change in the timber and wood products industries. Because technical change is the product of research, a review of harvesting history allows us to hypothesize where research has been productive and to anticipate where it may be productive in the future.

There was little incentive to develop land-saving technologies when land was relatively more plentiful than labor and where harvests originated from a ready inventory of mature timber. Timber production per acre actually decreased in the first half of the twentieth century—although aggregate output expanded during this time. (Output per acre decreased as stand density, access, species quality, etc. declined with the expanding frontier.) Incentives favored labor-saving technologies, which were *mechanical* (as opposed to biological) and which concentrated on wood harvesting, utilization, and processing (as opposed to timber growing) activities. In the future, as land becomes relatively scarcer, this picture may change considerably. There may be an expanding search for ways to increase production per acre through labor-saving technical changes which will be of a relatively more biological orientation than in the past.

The available evidence supports these hypotheses. The well-known advances in wood harvesting, utilization, and processing are numerous:

1. Introduction of steam-powered sawmills and steam-powered locomotives in harvesting (c. 1870)
2. Introduction of the band-saw, reducing loss to sawdust (c. 1870)
3. Introduction of gasoline and diesel powered mills and woods machinery (c. 1910, exclusive use by 1930)
4. Introduction of power saws for felling and bucking (c. 1920)
5. Introduction of the sulphate pulping process permitting the use of resinous southern pines (c. 1920)

6. Introduction of wheeled and crawler tractors for skidding (c. 1930)
7. Sawmill electrification (c. 1932)
8. Introduction of the semi-chemical pulping process permitting the use of hardwoods (c. 1940)
9. Development of the chip-n-saw, permitting expanded use of sawmill waste in pulping (c. 1962)
10. Development of improved lathes for plywood and veneer manufacture contributing to the opening of the southern pine plywood industry (c. 1962)
11. Introduction of computerized scanning in sawmills, improving log utilization (c. 1970)
12. Decreasing standards in dimension lumber (2 × 4's are now actually 1½ × 3½ inches)
13. Increasing mill specialization, permitting decreased standards in harvested logs (Acceptable logs in the West were 32 feet in length. They are now 16 feet. Eastern standards have decreased from 16 to 8 feet.)

While biological advances such as thinning, fertilizing, and various improvements on natural regeneration were known, they have been applied only rarely and haphazardly in North America until recently. Industrial forest landowners do not apply them universally even now and non-industrial owners often apply these timberstand improvements only if subsidized to do so.

Forest fire control may be the exception. Certainly it has consumed major public effort since the 1930s. Nevertheless, the impact of the fire control is difficult to demonstrate. Some timber that was protected from fire only burned later, some remains inaccessible or has never been harvested for other reasons, some only replaces other timber of marketable age and quality and, of course, some made a net contribution to social welfare. The proportion in the final category is incalculable with existing data.

The few quantitative analyses support these hypotheses. Horsepower per woods-worker has increased steadily since 1919 and the capital-labor ratio has similarly increased—at a rate more rapid than that in either paper or metal products. Nevertheless, a decline in land productivity from 1910 until 1950 offset it—causing a decline in labor productivity until 1940 (Ruttan and Callahan 1962). Perhaps the decade of the 1940s marks the beginning of a change from a frontier to a developed timber economy, but the value-added due to lumber processing continues to be relatively small even today and the rates of both productivity growth and technical progress in the lumber industry are approximately half of those for all industries for the United States in the twentieth century (Kendrick 1961, Stier 1980).

These rates have increased only somewhat since 1949 (Robinson 1975). Value-added by processing activities in the pulp and paper industry is much larger and the rates of both production and technical progress in this industry are more commensurate with those in other manufacturing industries. Moreover, there has been a bias toward labor-saving technical change in all wood products industries, but particularly in the sawmill and planing mill industry (Hunter 1955, Holland 1960, Stier and Bengston 1991).

EXPECTATIONS

We might anticipate that these trends will continue awhile longer, but without their previous strength. Forestry in general, and southern forestry in particular, is converting to more land-saving timber growing technologies. While biological research probably has not been widely adopted and, therefore, has not been highly productive, it may become more so. If we find that the economic returns on biological research have been low then we might expect to find only a small amount of private research investment. Indeed, in the past, the returns have been too uncertain and too distant to attract much private investment. The many small timber producers may have also found investments in tissue culture and thinning regimes, for example, a severe strain on their limited financial reserves. Furthermore, it is difficult for investors to establish proprietary claims to the return on these biological research investments. These are all reasons why, in the past, public agencies may have been responsible for the largest share of timber growth and management research. While the private industrial component of biological research may expand in the future, these are also reasons why some public research will continue to be important.[6]

There may be considerable transfer to forestry of new biological technologies from American agriculture, where there is more experience with biological growth functions, and from European forestry, where land has traditionally been a scarcer factor in timber production. Moreover, as the leading timber producing region, the South may expect its growth and management research to produce benefits which eventually spillover to other timber producing regions of the country. Research directed toward utilization and process technologies will continue to be attractive, as will labor-saving innovations in general. Biological research and technical change, however, may become more important components of the total forestry research effort. The subsequent empirical chapters should enlighten us regarding some of these hypotheses.

NOTES

1. Also see Evenson, Waggoner, and Ruttan (1979).

2. There have also been a number of detailed case studies of particular innovations which trace out subsequent consequences. Mansfield and his colleagues (1968, 1971, 1977) provide notable examples. Griliches (1973) surveys these.

3. Davis (1967), Porterfield (1974), Pee (1977), Thomson (1983), Lofgren (1988), McKenney et al. (1988), and Xiao (1988). There is work in progress at the University of Minnesota by H. Gregersen, D. Bengston and their colleagues and by J. Davis at the Australian Centre for International Agricultural Research.

4. See Spelter (1948) for regional lumber production data from 1799.

5. Policies requiring "even-flow" of annual harvest volumes and restricting harvests to timber aged one and one-half times its biological maturity (itself, often twice the economic maturity) are examples. Others can be found in the expanding Wilderness system, in state forest practice acts, in high logging road standards, and in restrictions on herbicide and insecticide use.

6. Callaham (1981) estimates that expenditures for forestry research and development were $331 million in fiscal year 1978. Thirty-six percent of this was US Forest Service, 6 percent other public, and 58 percent private. *Business Week's* (1979) estimate of private forestry research expenditure for the same year is a more modest $162.3 million.

The Benefits of Forest Products Research: A Dual Approach

OUR OBJECTIVES in the next three chapters are to be specific in our demonstration of the analytical approach yet sufficiently wide-ranging in our choice of industries to leave a general impression of the historical merits of research for forest products sector as a whole. We hope to accomplish both objectives by developing the approach in detail (chapter 3), then applying it with care for the softwood plywood (SWPW) industry (chapter 4), and finally surveying similar applications in three additional industries (chapter 5). Readers who are uninterested in the detailed analytical technique might focus on the written review at the beginning of the next section and forgo the subsequent mathematical detail later in this chapter.

Chapter 1 provided a basic discussion of the two general methods for research evaluation, the production function method and the consumers' and producers' surplus method. Each has its advantages but data deficiencies restrict use of the production function approach in forestry.[1] Therefore, the recent forestry literature features the consumers' surplus approach. It estimates increases in consumers' surplus, nets out research costs, and calculates an internal rate of return for R&D expenditures. Our approach in this chapter marries the two methods by recognizing that the production function is the dual of the supply function used for measuring consumers' and producers' surpluses.

It is a "dual" in economic jargon because there is a one-to-one relationship between these two functions and because the estimation procedure relies on (more available) input prices rather than input quantities to estimate both functions. (If input prices permit us to identify one point on one function, then they also permit us to identify the corresponding point on the other.) The dual has the advantages of both analytical methods. It permits direct estimates of the net economic benefit (as from the consumers' surplus approach) and the value of the marginal product (as from the production function approach) resulting from research.

Our approach also improves on previous estimation procedures by allowing the costs of other (non-research) factors of production to vary throughout the period of inquiry. Real factor costs (*e.g.*, wages) may vary substantially over periods long enough to incorporate the full benefits from a research breakthrough. Therefore, changes in non-research factor costs can themselves cause important shifts in industrial production and our incorporation of and adjustment for these changes improves the accuracy of estimated research benefits. For example, the geographic locus of the SWPW industry shifted from the Pacific Northwest in 1950 toward the South and Southeast by the 1980s—perhaps because of lower southern wages, as well as any research breakthrough enabling production from southern tree species. It would be a mistake to confuse production gains due to research with those due to adjustments in other factor payments. Therefore, this second analytical improvement may have particular importance for the empirical examination in chapter 4.

This chapter begins with a brief review of the approaches applied in other evaluations of forestry research before contrasting these approaches with a general discussion of our approach. This general discussion provides context for the detailed mathematical discussion that follows as the feature of this chapter. The mathematical detail proceeds from an initial explanation of the basic functions to the derivation of four fundamental conceptual results, each important to subsequent discussions of research investment and research policy: (1) the value of marginal product, (2) the marginal rate of return, (3) consumers' surplus and (4) the net economic benefit due to public research inputs. The final section reviews our analytical approach and summarizes its comparative advantages.

RECENT RESEARCH EVALUATIONS IN FORESTRY

Colleagues at the University of Minnesota have been the leaders in introducing impact evaluations of forestry research. Bengston's work is probably the best-known. Bengston (1983, 1984) provides two examples of impact evaluations in his examinations of research leading to the development of the structural particleboard (SPB) industry. SPB is a new product that substitutes for softwood plywood as sheathing material. Product innovation is difficult to evaluate. Bengston finesses this analytical problem by treating this new product as a new *process*. That is, supply before the SPB innovation is the supply of softwood plywood while the less-expensive supply after the innovation is supply of SPB. Thus, the R&D that created SPB reduced the marginal cost of sheathing and shifted the supply function for sheathing downward. The consumers' surplus associated with this downward shift measures the social benefit of SPB R&D.

The basic empirical assumptions necessary for this analysis are that the demand price elasticity for SWPW is a proxy for the demand price elasticity for SPB and that supply of both SWPW and SPB is perfectly elastic. The second assumption follows Griliches (1958). It generates a conservative estimate for consumers' surplus and implies that producers' surplus is equal to zero. Bengston also forecasts logistic growth in SPB demand until the year 2000.

Therefore, for any period t, the increase in consumers' surplus is

$$\Delta CS_t = (P_{pw,t} - P_{pb,t})\, Q_{pb,t}\, (1 - k_t n/2) \tag{3.1}$$

where ΔCS_t = change in consumer surplus
$P_{pw,t}$ = price of plywood sheathing
$P_{pb,t}$ = price of structural particleboard
$Q_{pb,t}$ = quantity of structural particleboard consumed
n = the price elasticity of demand for SWPW (a proxy for the SPB elasticity)
k_t = $(P_{pw,t} - P_{pb,t})/P_{pw,t}$

Equation (3.1) follows Griliches (1958), as also reported by Norton and Davis (1981).

The social benefit of SPB research is the measure of consumers' surplus minus the R&D costs. Reliable cost estimates are a problem. Bengston assembles an estimate of previous public research expenditures, E, beginning with two counts of screened and published SPB research, R, and knowledge of the average number of publications per scientist year, R/S, in order to estimate the input of government scientist year, $R(R/S)^{-1} = S$. He multiplies the resulting estimate by Callaham's (1981) estimate of US Forest Service costs per scientist year in 1977 and adjusts annually according to Sonka and Padberg's (1979) academic R&D price index. The result is an annual estimate of *public* (that is, US Forest Service) research costs. Industry estimates of *private* research costs complete the picture. The sums of public and private research cost estimates provide a total cost series for R&D for the years prior to 1981. Bengston projects future research costs to continue increasing at the 4.1 percent annual rate observed by Sonka and Padberg for the years 1947-1979.

Bengston's comparison of the change in consumers' surplus from equation (3.1) with the discounted sum of total public and private R&D costs yields an internal rate of return for SPB research between 18 and 22 percent depending on various assumptions imposed to check the sensitivity of the estimates.

Bengston's approach has appeal in its clear and straightforward design and in its modest data requirements. Bengston (1985) repeats his analysis for the aggregate of all lumber and wood products research. Three investigations by other analysts examine research in other forest industries. All draw on essentially the same consumers' surplus approach. Table 3.1 reports their results. All are in the neighborhood of returns observed for various research activities in agriculture by Ruttan (1980) and repeated in chapter 2 of this book.

The Dual Approach

Our alternative approach is more data intensive than Bengston's but it has the advantage of directly measuring the output effect of R&D expenditures. It also controls for factors other than research which might shift the supply function. We will first discuss this approach in general terms and with reference to figure 3.1, and then develop its full mathematical detail.

In general terms, we begin by specifying an industry production function that includes public research as one factor of production, continue by solving the profit maximization problem from this production function, and then derive the industry supply function. We can estimate this supply function simultaneously with demand. Many of the usual measures of research gains derive from the estimated supply function. In the final analysis, however, these measures must be net of some judgment about R&D costs.

This approach produces endogenous estimates of both the price elasticities of supply and demand and the output elasticities of the various productive factors, including the output elasticity of public research effort. The output elasticity of research traces through as a shift parameter in production and supply. Estimates of research benefits follow from exogenous information regarding changes in research effort and this endogenous knowledge of the research induced supply shifts. The resulting research benefits are "nonresidual" because the production function measures these benefits directly in the same manner that it measures the effects on supply of changes in any other particular productive factor.[2]

This procedure is conceptually similar to estimating the effects of an exogenous shock on any stable system. For a given period t, we have observations of price and quantity and estimates of the price elasticities of supply and demand and the output elasticity of research. Suppose, for example, that price and quantity observations and price elasticity estimates reflect demand curve D_t and supply curve S_t in figure 3.1. Suppose,

Table 3.1. Research evaluation results in forestry

author	forest product or industry	analytical approach	internal rate of return (%)[*]
Bengston (1984)	structural particleboard	consumers' surplus	18-22
Haygreen et al (1986)	lumber,plywood, pulp and paper	consumers' surplus	14-36
Westgate (1986)	containerized seedlings	consumers' surplus	37-111
Bengston (1985)	aggregate lumber and wood products	consumers' surplus	34-40
Bare and Loveless (1985)	forest nutrition	consumers' surplus	9-12

Chang (1986) considers returns to growth and yield research for loblolly pine. He does not report an IRR, but does calculate a benefit/cost ratio of 16:1.

[*]All estimates are for average, not marginal, internal rates of return.

furthermore, that today's R&D begins to impact production k periods in the future. (Current R&D impacts production only after the delays required to complete the research, to transfer the research findings to the factory, and then to retool the factory.) We can construct the supply curve S_{t+k} for period $t + k$ from knowledge of the observed level of research effort during period t and the output elasticity for research in period $t + k$. This is the supply curve expected in period $t + k$ with the known level of research effort in period t, but with no change since period t in the prices or quantities of other productive factors. (The mathematical formulae later in this section add sophistication by permitting changes in factor prices and quantities from period t to period $t + k$). Suppose that demand also remains unchanged between periods t and $t + k$. The social benefit in period $t + k$ of research conducted in period t is the area between the two supply curves and under the demand curve.

The impact of R&D conducted in period t does not end in period $t + k$ but usually carries over into the future. The lag structure of the production function and econometric estimates of supply and demand will suggest an output elasticity of research conducted in period t for supply in period $t + k + 1$. Therefore, figure 3.1 constructs the supply curve for period $t + k + 1$ as S_{t+k+1}. The social benefit in period $t + k + 1$ of research conducted in period t is the area between S_t and S_{t+k+1} and under the demand curve. Figure 3.1 shows a case where the benefits of R&D depreciate as new re-

Price

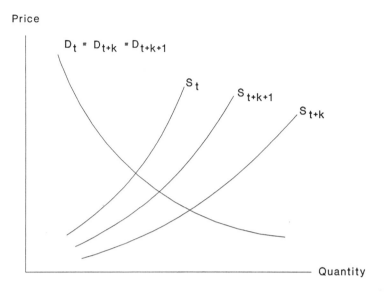

Figure 3.1. Future supply curves due to research conducted at time *t*

search and the new capital embodied in it replace old research and capital of a less recent vintage. Subsequent R&D impacts continue to depreciate and subsequent supply curves each shift again toward the original supply curve S_t until the difference between the original supply curve and the newest supply curve is arbitrarily small and we can ignore further benefits from R&D conducted in the original period *t*. The total accumulation of social benefits is the discounted sum of research-generated benefits for all periods from S_{t+k} until that period when the research impact fully depreciates and the supply curves for successive periods coincide.

Our imperfectly elastic supply curves mean that part of the research benefits from each period fall below that period's market equilibrium price in figure 3.1. These benefits accrue to producers and they are important for understanding producers' incentives to invest in research. Producers' surplus, however, is conceptually meaningful only for competitive or contestable markets. Forest products markets generally satisfy this condition.[3]

The final analysis of net research benefits also requires some understanding of R&D investment costs. Previous forestry research evaluations estimate the full return to all public and private research investments. Our policy focus in this volume concentrates on returns restricted to public research. Therefore, our estimate begins with a measure of public R&D

costs: the multiple of government scientist time and the cost of a unit of that time. Our unit cost construction is similar to Bengston's.

To the full public costs, we add a measure of the private implementation effort necessary before public research can have its impact on output. Public agencies fund the research and make the research results available but they stop short of funding the technical engineering modifications necessary to introduce the research results to the production line. These engineering modifications require private funding. These private implementation costs are distinct from both (a) private research costs and (b) the private implementation costs of private research. We use examples of private implementation costs for select public research breakthroughs to suggest a general cost ratio for private implementation of all public research investments. If this ratio is consistent across all public research expenditures, then the total of public research and its private implementation costs is the multiple of the public costs and this ratio.

Combining the measure of social benefit with the cost estimate for public research and its associated private implementation permits the derivation of both the net economic benefit of public research and the rate of return. Summing the aggregate of single period supply shifts due to increased research (and implementation) inputs and taking the derivative with respect to a unit of research input generates a measure of the value of the marginal research product. The results section of our subsequent mathematical formulation describes the derivation of these three measures in detail. This derivation is a straightforward extension of the production function approach to research evaluation.

Mathematical Formulation of the Dual

The first step in this more rigorous statement of our approach is choice of a technical form for the production function. The second and third steps have to do with converting production to supply, assessing demand, and simultaneously estimating the supply and demand functions in order to find the correctly specified production-supply dual.

Griliches (1979) suggests the Cobb-Douglas form for those research evaluations, like ours, that focus on output effects rather than interactions among productive inputs. For the softwood plywood case, production has been relatively capital intensive over the entire period of our inquiry (1950-1980) and there is little reason to expect radical shifts in either the capital or labor shares. Moreover, even the relatively simple Cobb-Douglas form sacrifices valuable degrees of freedom to the lags between research and implementation. More complex forms require the estimation of additional coefficients and, therefore, sacrifice even more degrees of freedom.[4]

Following Griliches suggestion, the production function at time t is[5]

$$Q_t = L_t^{\alpha_1} K_t^{\alpha_2} Z_t \tag{3.2}$$

$$\text{where } Z_t = \prod_{i=i_0}^{\infty} \left(H_{t-i}^{\eta} \, G_{t-i-j_0}^{\mu} \right)^{\lambda^{i-i_0}}; \qquad i_0 > 0, \, j_0 \geq 0 \tag{3.3}$$

is an index of technology and

Q_t = total quantity produced
L_t = labor services
K_t = capital services
Z_t = index of technology
H_{t-i} = private R&D effort, $i = i_0, \ldots, \infty$
G_{t-i-j} = government R&D effort, $j = j_0, \ldots, \infty$

Beginning with equation (3.2) and substituting from equation (3.3), yields the final form of the production function.

$$Q_t = L_t^{\alpha_1} K_t^{\alpha_2} \prod_{i=i_0}^{\infty} \left(H_{t-i}^{\eta} \, G_{t-i-j_0}^{\mu} \right)^{\lambda^{i-j_0}} \tag{3.4}$$

The lag structure in equation (3.3) bears closer examination. The lags attributed to the two R&D variables indicate the delay between the initial research investment and the eventual implementation of research results. Publicly funded R&D may be subject to lengthier lags than private R&D. Pakes and Shankerman (1984) provide important evidence of the general lag. Griliches (1979) supports the contention that public lags are greater, with the argument that public research is generally more basic and private research more applied; therefore, public research must go through additional time-consuming phases before its implementation.

Research lags are relatively continuous in the forest products industries because the stream of new techniques is relatively continuous and because research information spreads rapidly among the limited number of firms found in each industry. Moreover, the forest products industries can often adapt new techniques to existing machinery, thereby avoiding large and discontinuous injections of implementation expenditures. In the SWPW industry, for example, many important research breakthroughs required only small changes in operations (*e.g.*, glue substitutes, lathe adjustments, modifications in log heating temperatures). Furthermore, users tend to learn of technical advances quickly since private firms maintain close contact with the Forest Products Laboratory, the primary source of public research.

The lag structure in equation (3.3) is the familiar Koyck, or geometric, form. The Koyck lag implies that the larger productivity effects of research occur initially, after the adoption period i_0, and that these effects decay exponentially over time as new technologies replace old technologies. Other lag structures, such as the inverted V (research impacts on output begin small, rise, reach a maximum, and then fall), may be more appropriate for other industries in which the application of research requires large initial capital investments (therefore, firms have an incentive to depreciate existing capital before replacing it with new technologies) or in which dispersal of information about new technologies occurs gradually over a long period of time.[6]

The expenditure term for private R&D also bears closer examination. Firms fund their research from a portion of total revenues. Therefore, firms set

$$H_{t-i} = f P_{t-i} Q_{t-i} \qquad (3.5)$$

where P_t = real price per unit of output and f = the fraction of total revenues spent on private R&D. The term f is constant over the period of inquiry.

There are numerous empirical observations of constant proportional funding for research investments (Mansfield 1968, p. 62; Bullard and Straka 1987; NSF 1981; and annual reports of R&D expenditures in *Business Week*). Appendix 3A reviews some of the evidence. Furthermore, Seldon (1985, pp. 50-59) shows that this is the optimal behavior for a competitive firm employing Cobb-Douglas technology in a dynamic setting and under conditions of steady growth. The firm's optimal level of investment in R&D is independent of the firm's (and the industry's) optimal static choice regarding the employment of other productive inputs, so long as there are no adjustment costs for changes in the levels of employment of the various inputs.

Supply

If competitive firms maximize their cash flows for each period and finance R&D from these cash flows, then the industry maximizes net revenues in each time period:

$$\max_{L_t,\, K_t} \Pi_t = P_t Q_t - W_t L_t - R_t K_t - H_t \qquad (3.6)$$

where Π_t = industry profit
   W_t = real hourly wage
   R_t = real user cost of capital

All other variables were previously defined. The level of current private research investments H_t is not a choice variable in the instantaneous profit maximization decision because H_t does not affect the level of current total revenues. Rather, private research investments in period $t - i$ are a determinant of output in period t. [See equation (3.4).]

The supply function $Q_{S,t}$ derives from the maximization of this general profit function using equations (3.4) and (3.5) and Hotelling's lemma to solve for the quantity supplied. Appendix 3B shows the solution in detail. The estimated form of the supply function is

$$
\begin{aligned}
q_{S,t} = {} & (1 - \lambda)\ln A + \gamma(\alpha_1 + \alpha_2)(p_t - \lambda p_{t-1}) - \gamma\alpha_1(w_t - \lambda w_{t-1}) \\
& -\gamma\alpha_2(r_t - \lambda r_{t-1}) + [(\lambda - 1)\ln(1 - f)] + \gamma\eta h_{t-i_0} \qquad (3.7) \\
& + \gamma\mu g_{t-i_0-j_0} + \lambda q_{t-1} + \epsilon_{S,t}
\end{aligned}
$$

where $\ln A$ and γ are constant functions of α_1 and α_2 and $\epsilon_{S,t}$ is a normally distributed random error term. Lower case letters represent logarithmically transformed capitalized Roman letters.[7]

The constant term in brackets, composed of a function of the distributed lag term λ and the proportion of firm revenues spent on research f, is small for reasonable estimates of λ and f.[8] We can disregard it. This also means that $\ln(P_{t-i}Q_{t-i})$ can act as a proxy for $\gamma\eta h_{t-i}$ and the term in brackets.

This supply function is the focus of our econometric assessment of softwood plywood research in chapter 4. It is also the dual of the production function. It contains all the coefficients of the original production function defined in equation (3.4). Therefore, it can form the basis for our later derivation of the value of the marginal product of public research later in this chapter.

Demand

Most forest products are intermediate products. Their demands derive from the supplies of final products (such as residential construction, the final product associated with the largest share of softwood plywood consumption). The procedure for deriving the demand function, although simpler in method, is completely analogous to our earlier derivation of the supply function. It is simpler because there is no need to account for the affect of technical change on demand other than through changes in output and in the prices of substitutes and complements in production of the final product. The industry profit function for the final product appears in the same form as equation (3.6).

Maximizing the industry's profit function, and solving for quantity using Hotelling's lemma, provides a log linear derived demand equation expressed in current period prices.

$$q_{D,t} = \beta_0 + \beta_1 p_t + \beta_2 w_{d,t} + \beta_3 r_{d,t} + \beta_4 p_{l,t} + \beta_5 p_{c,t} + \epsilon_{D,t} \qquad (3.8)$$

where $q_{D,t}$ = quantity of wood product demanded
$w_{d,t}$ = real wage in the downstream industry
$r_{d,t}$ = real user cost of capital in the downstream industry
$p_{l,t}$ = real substitute input (lumber) price index
$p_{c,t}$ = real price per square foot of new residential construction
$\epsilon_{D,t}$ = a random and normally distributed error term

The market clearing assumption, $q_{D,t} = q_{S,t}$, ensures a solution for the system of equations (3.7) and (3.8). Simultaneous estimation of the supply and demand system requires nonlinear methods because the supply equation (3.7), even excluding the bracketed term, is nonlinear in its coefficients. Our subsequent analysis uses the demand parameters to determine the own price elasticity and the supply parameters to determine the own price and the public research elasticities.

This general system of supply and demand equations differs from others found in the forestry literature. The inclusion of an endogenously determined supply shifter as well as the constrained form of the supply function differentiate the system from spatial equilibrium supply models such as TAMM (Adams and Haynes 1980). Both the simultaneous estimation of factor coefficients and the Cobb-Douglas technology, among other things, differentiate the system from translog factor demand models (*e.g.*, Stier 1980, Nautiyal and Singh 1985). Finally, simultaneous estimation and the inclusion of research inputs as implicit shifters in product supply differentiate this system from Spelter's (1985) examination of the effect of technological diffusion on lumber demand, as well as from previous research evaluations in forestry and elsewhere.

RESULTS

Three sets of results derive from this model: the value of the marginal product for public research, the internal rate of return on public research and various measures of the economic benefit of public research. We divide the value of the marginal product into its short- and the long-run measures and derive it first.

The Value of the Marginal Product, and the Marginal Internal Rate of Return

The value of the marginal product (*VMP*) measures the addition to output originating from the last unit of input, the last unit of public research in our case. For Cobb-Douglas technologies, the *VMP* is the value of the average product times its associated exponent. Therefore, if we assume an immediate impact of R&D on output, then the short-run (or initial period) *VMP* is:

$$VMP_{rd,t}^{SR} = P_t(\partial Q_t / \partial E_t) = \mu P_t Q_t / E_t \tag{3.9}$$

where E is the dollar value of social investment in public R&D. (Further explanation of E follows completion of this discussion of marginal returns.) Given the general research-implementation lag structure and, particularly, the two-period lag ($i_0 = 2$, $j_0 = 0$) which we will observe in the SWPW case, then the more accurate specification of the short-run *VMP* is

$$VMP_{rd,t}^{SR} = \mu P_{t+2} Q_{t+2} / (1 + \rho)^2 E_t \tag{3.10}$$

where ρ is the assigned social discount rate.

This *VMP*^{SR} contrasts with the long-run *VMP* which measures the sum of continuing periodic output affects due to a one time marginal public research input. The Koyck lag structure in equation (3.3) and our chapter 4 observation of the SWPW case suggest that output increases begin in the second period and continue, if at decreasing levels, for an extended period of time. Successive periodic marginal products can be expressed as:

$$\partial Q_{t+2} / \partial E_t = \mu Q_{t+2} / E_t$$
$$\partial Q_{t+3} / \partial E_t = \lambda \mu Q_{t+3} / E_t$$
$$\cdot$$
$$\cdot$$
$$\cdot$$
$$\partial Q_{t+m} / \partial E_t = \lambda^{m-2} \mu Q_{t+m} / E_t$$

and the long-run *VMP* is

$$VMP_{rd,t}^{LR} = (\mu/E_t) \sum_{i=2}^{\infty} \left[\lambda^{i-2} P_{t+i} Q_{t+i} / (1 + \rho)^i \right] \qquad (3.11)$$

Estimating equation (3.11) directly requires price and quantity forecasts. We can simplify the task with a myopic version of market adjustment. (This is in the spirit of the usual calculation of the marginal internal rate of return.) The market value of output is $P_t Q_t$ at the time the research is undertaken. The price-quantity relationship assumes this value for each period—instead of the stream of unknown future market values. This simplifies equation (3.11) to

$$\begin{aligned} VMP_{rd,t}^{LR} &= (\mu P_t Q_t / E_t) \sum_{i=2}^{\infty} \left[\lambda^{i-2} / (1 + \rho)^i \right] \\ &= \mu P_t Q_t (1 + \rho) / (1 + \rho - \lambda)(1 + \rho)^2 E_t \end{aligned} \qquad (3.12)$$

The only difference between equations (3.10) and (3.12), the short-run and long-run *VMP*s, respectively, is that the long-run *VMP* includes the effect of the lag coefficient λ. The two equations are equivalent where the initial research impact is the only output effect. In this case there are no subsequent research effects and $\lambda = 0$.[9] As the effects of research output extend over time, λ increases and VMP^{LR} diverges from VMP^{SR}.

Since equation (3.12) generates a different return for each period, it is also common to report an average *VMP* for the periods in the sample. We follow Griliches (1964) and Peterson (1967) in reporting the geometric mean.

$$VMP_{GM}^{LR} = \left(\prod_{t=1}^{n} VMP_{rd,t}^{LR} \right)^{1/n} \qquad (3.13)$$

where n is the number of periods in the sample.

The notion of a long-run *VMP* for research is unusual in the research evaluation literature unless the marginal internal rate of return (*MIRR*) is also reported. The *MIRR* is the return on capital that makes the accumulated stream of all periodic benefits due to the final public research inputs just equal in value to that final research input. Or, for a final research input of one dollar, the *MIRR* is the ρ that equates the mean long-run *VMP* from equation (3.12) with unity.

The literature is not consistent in its computation of the *MIRR*. Our calculation of the *MIRR* is a special case (due to the lag structure) of the most common calculation. It is the value of ρ such that [in Davis' (1981) notation]

$$VMP_{GM}^{LR} \left[\prod_{i=1}^{n} W_i/(1 + \rho)^i \right]^{1/n} = 1 \qquad (3.14)$$

[For SWPW, $W_1 = 0$ and $W_i = \lambda^{i-2}$ for $i > 1$ and the first term of equation (3.14) is simply the (average) *VMP* defined by equation (3.13).]

Finally, we postponed until now the detailed calculation of research expenditures E_t. Both the *VMP* and the *MIRR* require an accounting for research expenditures, yet we have no direct measure of these expenditures. Our alternative builds on a measure of government scientist weeks G_t. The derivation of the actual *VMP* uses the identity that for any time period m

$$
\begin{aligned}
P_{t+m}(\partial Q_{t+m}/\partial E_t) &= P_{t+m}(\partial Q_{t+m}/\partial G_t)(\partial G_t/\partial E_t) \\
&= (\lambda^{m-i_0-j_0}\mu P_{t+m}Q_{t+m}/G_t)(\partial G_t/\partial E_t)
\end{aligned}
$$

An assumption that the relation of scientist weeks to research expenditures $(\partial G_t/\partial E_t)$ is constant for all levels of G_t permits estimation of the second right-hand equality. The term $\partial G/\partial E$ is the inverse of $\partial E/\partial G$ or, in other words, the inverse of the social costs of a scientist week C. Since $C_t G_t$ is the total social cost of research, the definition can be rewritten as

$$P_{t+m}(\partial Q_{t+m}/\partial E_t) = (\lambda^{m-i_0-j_0}\mu P_{t+m}Q_{t+m}/C_t G_t) \qquad (3.15)$$

Estimates of the costs of a scientist week are available. The SWPW analysis in chapter 4 uses equation (3.15) to calculate the *VMP* and *MIRR* estimates.

Total Economic Benefit

The calculations of changes in consumers' surplus and total economic benefit are more complex. There are three steps. The first is to define supply and demand for each period affected by an initial research investment. The second is to determine the increase in consumers' surplus attributed to the initial research investment in each of those periods. The third step accumulates the increases in consumers' surplus from all relevant periods. Calculations of producers' surplus and total economic benefit proceed analogously.

We begin by referring to the estimated form of the supply equation (3.7) and let a_1 and a_2 be the price elasticity of supply and the public R&D elasticity of supply, respectively.

$$a_1 = \gamma(\alpha_1 + \alpha_2)$$
$$a_2 = \gamma\mu.$$

where $\gamma = (1 - \alpha_1 - \alpha_2)^{-1}$, as defined in appendix 3B.

Research at time t begins to impact supply in period $t + i_0 + j_0$, or after the public and private research and implementation lags. The anticipated future supply equations are

$$q_{t+i_0+j_0} = F_t^S + a_1 p_{t+i_0+j_0} + a_2 g_t$$
$$q_{t+i_0+j_0+1} = F_t^S + a_1 p_{t+i_0+j_0+1} + \lambda a_2 g_t$$
$$\vdots$$
$$q_{t+i_0+j_0+k} = F_t^S + a_1 p_{t+i_0+j_0+k} + \lambda^k a_2 g_t$$
$$\vdots$$

where F_t^S is the intercept of the supply curve in period t. In general

$$Q_{t+i_0+j_0+k} = A_t^S G_t^{\lambda^k a_2} P_{t+i_0+j_0+k}^{a_1}; \qquad k = 0, 1, 2, \ldots \qquad (3.16)$$

where A_t^S is the antilogarithm of F_t^S.

Similarly, expected future demand is

$$Q_{t+i_0+j_0+k} = A_t^D P_{t+i_0+j_0+k}^{-\beta_1}; \qquad k = 0, 1, 2, \ldots \qquad (3.17)$$

where $-\beta_1$ is the demand price elasticity.

Setting supply equal to demand and solving for the anticipated market clearing price yields

$$P_{t+i_0+j_0+k}^e = (A_t^D / A_t^S)^\delta \, G_t^{-a_2\delta\lambda^k} = P_t G_t^{-a_2\delta\lambda^k}$$

where $\delta = (\beta_1 + a_1)^{-1}$. The final term originates from the solution for price at time t in the original supply and demand system,

$$Q_t = A_t^S P_t^{a_1} \tag{3.18}$$

and

$$Q_t = A_t^D P_t^{-\beta_1} \tag{3.19}$$

The change in consumers' surplus in period $t + i_0 + j_0 + k$ due to government R&D in time t is equal to the area under the demand curve and between the market equilibrating prices for periods t and $t + i_0 + j_0 + k$:

$$\int_{P_{t+i_0+j_0+k}^e}^{P_t} A_t^D P_t^{-\beta_1} dP = A_t^D (1 - \beta_1)^{-1} (1 - G^{\xi \lambda^k}) P_t^{(1-\beta_1)} \tag{3.20}$$

where $\xi = -a_2(1 - \beta_1)\delta$. Therefore, the discounted value of the change in consumers' surplus in all future periods attributed to research conducted at time t is

$$PV_t^{CS} = \sum_{i=i_0+j_0}^{\infty} (1 + \rho)^{-i} (1 - \beta_1)^{-1} A_t^D P_t^{(1-\beta_1)} \left(1 - G_t^{\xi \lambda^{i-i_0-j_0}}\right)$$

$$= (1 - \beta_1)^{-1} P_t Q_t \sum_{i=i_0+j_0}^{\infty} (1 + \rho)^{-i} \left(1 - G_t^{\xi \lambda^{i-i_0-j_0}}\right) \tag{3.21}$$

This summation may be approximated by the finite number of time periods before the research impact decays and future supply is within some small neighborhood of the original function.

It is a simple step to subtract public research expenditures from equation (3.21) and to calculate the net present value $NPV_t^{CS} = PV_t^{CS} - E_t$ where E_t is R&D expenditure in period t. We call this the single period or short run net present value.

Most research is a multiple year activity. The multiple year or long run net present value is

$$NPV^{CS} = \sum_{n=0}^{N-1} (1 + \rho)^{-n} (PV_n^{CS} - E_n) \tag{3.22}$$

Solving for the ρ that sets equation (3.22) equal to zero yields the internal rate of return IRR^{CS}.

The calculation of producers' surplus proceeds similarly. The research induced shift in supply creates a gross gain to producers (the area between the two supply curves and below the equilibrium price) and a gross loss to producers (a transfer to consumers' surplus caused by the decrease in equilibrium price). The net change in producers' surplus in time period $t + i_0 + j_0 + k$ due to public research conducted in time t is

$$\int_0^{P^e_{t+i_0+j_0+k}} A_t^S (G_t^{a_2\lambda^k} - 1) P_t^{a_1} dP - \int_{P^e_{t+i_0+j_0+k}}^{P_t} A_t^S P_t^{a_1} dP$$

$$= A_t^S (1 + a_1)^{-1} (G_t^{\psi\lambda^k} - 1) P_t^{1+a_1} \tag{3.23}$$

where $\psi = a_2(\beta_1 - 1)\delta$. The accumulated and discounted return to producers from research at time t is

$$PV_t^{PS} = \sum_{i=i_0+j_0}^{\infty} (1 + \rho)^{-i}(1 + a_1)^{-1} A_t^S P_t^{(1+a_1)} (G_t^{\psi\lambda^{i-i_0-j_0}} - 1)$$

$$= (1 + a_1)^{-1} P_t Q_t \sum_{i=i_0+j_0}^{\infty} (1 + \rho)^{-i} (G_t^{\psi\lambda^{i-i_0-j_0}} - 1) \tag{3.24}$$

where the second equality derives from equation (3.18).

Once more, the summation of single period surpluses yields the multi-period surplus. Finally, calculating the single period and multi-period net present values in the terms of producers' surplus NPV_t^{PS} and NPV^{PS} is analogous to calculating the net present values in terms of consumers' surplus NPV_t^{CS} and NPV^{CS} in equation (3.22).

Total economic benefit is the sum of consumers' and producers' surpluses, equation (3.21) plus equation (3.24). Short run net economic benefit is the sum of the consumers' and producers' surpluses minus the research expenditure.

$$NPV_t^{NEB} = PV_t^{CS} + PV_t^{PS} - E_t \tag{3.25}$$

Net economic benefit for the entire stream of research gains (NPV^{NEB}) is the discounted sum of single research period producers' and consumers' surpluses net of expenditures. The internal rate of return for net economic benefits (IRR^{NEB}) is the value of ρ that sets equation (3.26) equal to zero.

$$NPV^{NEB} = \sum_{n=0}^{N-1} (1 + \rho)^{-n} (PV_n^{CS} + PV_n^{PS} - E_n) \qquad (3.26)$$

SUMMARY

This chapter explains the analytical approach which we subsequently apply for our research evaluations in the forest products industries. The approach begins with industry production as a function of research expenditures, both public and private, and the costs of other productive inputs, *e.g.*, labor and capital. The effects of the research inputs are different from the effects of other inputs. Most inputs affect production immediately. Research affects production only in some subsequent period, after the research expenditure and the technical research breakthrough resulting from it, the dissemination of information about this breakthrough and, finally, the implementation of production line modifications necessary to accommodate it. Once the results from a single research effort are known and implemented, however, the research impact on production continues over an extended period. The lags identifying (1) the period between initial research expenditures and final implementation and (2) the duration of impact, once implemented, are empirical questions the answers to which will vary from case to case.

Knowledge of the production function, including its critical lags, permits estimates of changes in output as a function of changes in research effort or changes in the levels of other inputs. Our production function is additionally complex because it must separate public and private R&D inputs in order that we may later address policy questions regarding only the public input. Transforming the known production function into a supply function (where output is now a function of product price) and simultaneously estimating the demand function permits an understanding of the output market impacts due to changes in any factor input, including public research inputs. Since research continues to affect output over time, a succession of periodic (perhaps annual) measures of the production impacts on supply and demand are necessary in order to show the continuing effect of a single research injection.

Adjustments for changes in other, non-research input costs over time may be necessary. For example, rising wages, taken alone, lead to higher final product prices and a smaller equilibrium level of market output (an upward supply shift in figure 3.1). An offsetting research breakthrough might decrease production costs to their former level (before the wage increase) and expand equilibrium market production of the final product to its former level. Without separating the wage and research effects on production, the research effect and the total effect would appear to be the same—zero. An advantage of our approach to measuring research impacts is that it separates the increased wage effect from the research break-through effect. It continues to make this separation even if wages change several times during the continuing impact of a single research break-through.

There are various alternative ways to measure the market impact of research investments. The technical detail of this chapter shows the calculations of (1a) changes in the output value for a unit change in research for both the period of initial research impact and over the full duration of output effects caused by a single research injection, and (1b) the rate of return on the marginal input.

After adjusting for input costs, our technical calculations of research impacts continue with a second set of calculations that measure (2a) the net changes in producers' and consumers' surpluses, either for a single year or for the accumulation of years affected by the single research investment, and (2b) the average rate of accumulated return on the research investment.

The technical discussion in this chapter is powerful. It extends the capabilities of the consumers' surplus and production function approaches appearing in the previous research evaluation literature. It also permits more accurate measures of results because it permits adjustments in the costs of other productive factors through time. These are real conceptual improvements and they should yield more realistic results.

The problem with this approach lies in the extensive demands it places on data. Even these data demands, however, may reflect an improvement on the production function approach to research evaluation. The usual production function approach requires data on input quantities. Our approach substitutes data on input prices. The latter are more readily available. They also incorporate adjustments for quality differences—an additional problem in lengthy time series quantity data.

The next two chapters display empirical applications of our approach, first for research investments in the softwood plywood industry and then, more briefly, in three other forest products industries. These applications should clarify points about the approach which may be difficult to understand in the purely conceptual discussion of this chapter.

APPENDIX 3A: TOTAL REVENUE AS A PROXY FOR PRIVATE R&D

This appendix reviews the empirical justifications for using total revenue as a proxy for private R&D expenditures. Data for private expenditures on R&D are unavailable for firms whose main product is softwood plywood. Data do exist, however, for many firms in other industries and for industry aggregates corresponding to many of the Bureau of Commerce Standard Industrial Classification (SIC) codes, including SIC 24, 25 and 26 (lumber and wood products, furniture, paper and allied products). We can search this evidence for apparent rules-of-thumb regarding R&D investments on the part of these firms or industries.

Of course, any rule set by management can be changed by management—perhaps in response to tax incentives, changes in management time preference, or changes in the scientific base of the industry. Nevertheless, any rule is probably stable over an extended period between such changes. Therefore, we may observe a stable share of revenues allocated to R&D for a few years and then observe a change to a new stable share which lasts another few years, and so on. Mansfield (1968, ch. 3) and Hay and Morris (1979, p. 446) report stable rules-of-thumb for R&D investments by firms. Grabowski (1970) and Nerlove and Arrow (1962) also mention such rules.

Seldon's (1985, pp. 49-59) empirical evidence for various firms in ten diverse industries from 1975 through 1980 extends our confidence. Seldon's industries were automotive parts and equipment, building materials, conglomerates, containers, food, instruments (measuring devices and controls), leisure time, miscellaneous manufacturing, paper, and steel. The evidence from this sample suggests that rule-of-thumb investment in R&D is widespread both for individual firms and among diverse industries, including the lumber and wood products and furniture aggregate which includes the softwood plywood industry.

Over his five-year sample period, Seldon rarely finds an annual change in the percentage of total revenue allocated to R&D in excess of 0.3 percentage points except where a discrete jump to some new percentage level occurs. Even then, the change remained small, usually less than two percentage points. Indeed, the stability of research shares becomes even more striking when we consider the range of firm expenditures for R&D over the entire period. We anticipate discrete jumps in research shares; therefore, large ranges in percentage shares for some firms. Yet of 123 firms, only 31 had ranges of 1.0 percentage point or greater for the full five years, while 73 had ranges of 0.5 percentage point or less and 21 had ranges of 0.1 percentage point or less. Of course, the magnitudes were small to start with, but this evidence does lend confidence to the notion that *firms* spend a relatively constant fraction of total revenue on R&D.

We might inquire whether changes in the share of total revenues invested in R&D show greater variation for the *industry* than for the firm. Table 3A.1 provides the evidence for annual observations from fourteen industries over the period 1967-1980. These industries include the forest products industries; SIC 24, 25, 26. Only three of fourteen industries have ranges greater than one percentage point for the entire thirteen year period. Only fifteen of 193 total observations exceed 0.2 percentage points of difference for any two consecutive years.

The forest product industries show ranges of R&D expenditures to total revenues not exceeding a change of 0.5 percentage points for the entire period and only once exceeding 0.1 percentage point difference between any two consecutive years. Thus, the rule-of-thumb seems as stable for industry-wide observations as for Seldon's observations of firms. This suggests that changes in research shares across firms within an industry somewhat offset each other and a rule-of-thumb assumption may even more closely approximate reality at the more aggregate industry level.

This all encourages us to accept total revenue as a proxy for private R&D in the industry. If we knew the years where discrete shifts occurred in the fraction of the total revenue spent on R&D, then we would have an even closer proxy. Such shifts, however, are difficult to identify with confidence. They are commonly due to accounting practices such as new tax incentives which encourage firms to label certain previously disallowed expenditures as R&D investments.

APPENDIX 3B: DERIVATION OF THE SUPPLY FUNCTION

The production function is

$$Q_t = L_t^{\alpha_1} K_t^{\alpha_2} Z_t \tag{3.2}$$

where $\quad Z_t = \prod_{i=i_0}^{\infty} (H_{t-i}^{\eta} G_{t-i-j_0}^{\mu})^{\lambda^{i-i_0}}; \qquad i_0 > 0, j_0 \geq 0 \tag{3.3}$

Firms (and the industry) maximize cash flow each period. Therefore, the industry solves the problem

$$\max_{L_t, K_t} \Pi_t = P_t Q_t - W_t L_t - R_t K_t \tag{3A.1}$$

Table 3A.1. Private R&D expenditures as a percent of total revenue in selected two-digit SIC code industries, 1967-1980

SIC code	industry	1967	1968	1969	1970	1971	1972	1973	1974	1975	1976	1977	1978	1979	1980	difference 1967-80
20	food and kindred products	.5	.5	.4	.5	.5	.4	.4	.4	.4	.4	.4	.4	.4	.4	.1
22, 23	textiles & apparel	.5	.5	.6	.5	.5	.4	.4	.4	.4	.4	.4	.4	.4	.4	.2
24, 25	lumber, wood products, furniture	.3	.4	.4	.8	.7	.8	.7	.8	.7	.7	.8	.7	.7	.8	.5
26	paper & allied products	.9	.9	1.0	.9	.9	.8	.7	.8	.9	1.0	.9	1.0	1.0	1.1	.4
28	chemicals & allied products	4.6	3.8	3.9	3.9	3.7	3.6	3.5	3.5	3.7	3.7	3.7	3.6	3.4	3.5	1.2
29	petroleum refining and related industries	.8	.8	.9	1.0	.9	.8	.7	.6	.7	.6	.7	.8	.7	.6	.4
30	rubber products	1.9	2.1	2.2	2.3	2.2	2.6	2.6	2.5	2.5	2.4	2.1	1.9	1.9	2.2	.7
32	stone, clay and glass products	1.8	1.6	1.7	1.8	1.8	1.7	1.7	1.7	1.2	1.2	1.2	1.1	1.1	1.3	.2
33	primary metals	.8	.8	.8	.8	.8	.7	.7	.6	.8	.8	.7	.6	.6	.6	.2
34	fabricated metal products	1.3	1.3	1.2	1.2	1.2	1.1	1.2	1.2	1.2	1.2	1.2	1.1	1.1	1.3	.5
35	machinery	4.2	4.0	3.8	4.0	4.0	4.3	4.3	4.6	4.8	4.9	5.1	5.1	5.1	5.6	1.8
36	electrical equipment	8.6	8.4	7.9	7.3	7.2	7.1	6.9	6.6	6.5	6.7	6.2	6.3	6.4	6.5	2.4
38	professional and scientific instruments	5.4	6.5	6.4	5.7	5.7	5.9	6.1	6.1	5.9	6.2	6.1	6.1	6.2	6.0	1.8
21, 27	other manufacturing	-	-	-	.8	.8	.8	.8	.9	.8	.7	.7	.7	.7	.9	.2

Source: National Science Foundation (1981)

subject to equations (3.2) and (3.3). The industry funds current R&D (H_t) from this cash flow—but H_t is not a control variable for this problem because R&D expenditures raise *future*, not current, cash flow.

Combining terms, taking first derivatives with respect to the control variables, setting these derivatives equal to zero and solving the new equations simultaneously yields

$$L^* = \alpha_1^{\gamma(1-\alpha_2)} \alpha_2^{\gamma\alpha_2} (PZ)^\gamma R^{-\gamma\alpha_2} W^{-\gamma(1-\alpha_2)} \tag{3A.2}$$

$$\text{and } K^* = \alpha_1^{\gamma\alpha_1} \alpha_2^{\gamma(1-\alpha_1)} (PZ)^\gamma R^{-\gamma(1-\alpha_1)} W^{-\gamma\alpha_2} \tag{3A.3}$$

$$\text{where } \gamma = (1 - \alpha_1 - \alpha_2)^{-1} \tag{3A.4}$$

Asterisks indicate optimal levels. Time subscripts are suppressed for convenience.

The profit in each period, after funding R&D, is

$$\Pi_t = P_t Q_t^* - W_t L_t^* - R_t K_t^* - H_t \tag{3A.5}$$

where Q_t^* is given by equation (3.2) subject to equations (3A.2)-(3A.4). Private expenditure on R&D in time t is

$$H_t = f P_t Q_t \tag{3A.6}$$

Substituting equations (3.1), (3.2), (3A.2)-(3A.4) and (3A.6) into equation (3A.5), equating Π_t with zero, and rearranging terms yields the supply function

$$Q_t = (1 - f)^{-1} A P_t^{\gamma(\alpha_1+\alpha_2)} W_t^{-\gamma\alpha_1} R_t^{-\gamma\alpha_2} \prod_{i=i_0}^{\infty} \left(H_{t-1}^{\gamma\eta\lambda^{i-i_0}} G_{t-i-j_0}^{\gamma\mu\lambda^{i-i_0}} \right) \tag{3A.7}$$

$$\text{where } A = \alpha_1^{\gamma\alpha_1} \alpha_2^{\gamma(1-\alpha_1)} + \alpha_1^{\gamma(1-\alpha_2)} \alpha_2^{\gamma\alpha_2} \tag{3A.8}$$

Taking the log of equation (3A.7) and subtracting $\lambda \ln Q_{t-1}$ yields a Koyck transformation which is free of the infinite lag

$$q_t = (1 - \lambda) \ln A + \gamma(\alpha_1 + \alpha_2)(p_t - \lambda p_{t-1}) - \gamma\alpha_1(w_t - \lambda w_{t-1})$$
$$- \gamma\alpha_2(r_t - \lambda r_{t-1}) + [(\lambda - 1)\ln(1 - f)] + \gamma\eta h_{t-i_0} \qquad (3A.9)$$
$$+ \gamma\mu g_{t-i_0-j_0} + \lambda q_{t-1} + \epsilon_{S,t}$$

where lower case letters replacing capitalized Roman letters represent logarithms, and γ and A are defined by equations (3A.4) and (3A.8), respectively. Equation (3A.9) is the same as the supply equation (3.6) in the body of the chapter.

NOTES

1. There are three data problems with the production function approach: (1) Aggregate measures of factor services (R&D, labor, and capital services) are often unavailable. (Analysts often compensate by using variables like levels of employment and capital stocks as proxies for these aggregate measures.) (2) Collinearity often exists in the factor service data or their proxies. This hinders statistical estimation of their coefficients. (3) Both sides of the input markets may use the productive factors, thereby causing simultaneity bias in the estimated coefficients (Griliches 1979).

2. In contrast, the Bengston-Griliches estimates (and all previous forestry research evaluations) produce no estimate for producers' surplus. Other, more detailed, approaches in the agriculture literature use exogenous estimates of the supply price elasticity. The omission of research inputs as productive factors or supply shifters in the (Bengston-Griliches) exogenous estimates creates indeterminate biases in the eventual measures of research impacts.

3. Product homogeneity and low entry costs are preconditions for classifying an industry as competitive. Forest products are often either homogeneous by nature or else standardized by government requirements or by industry councils. Entry costs are often low (with exceptions such as particleboard and paper).

The usual rule is to consider an industry competitive if its four largest firms share 40-50 percent of the market or less (Scherer 1980, p. 67). All forest products industries satisfy this rule. The largest four-firm concentration ratio for a four-digit SIC Code forest industry in 1977 was 48 percent [for both particleboard (SIC Code 2492) and pulpmills (SIC Code 2611)]. No others exceeded 38 percent. The average for all 23 four-digit forest product industries was 23.7 percent.

4. Evenson (1981) suggests a more general form where the link between past research and current output is less important. Griliches and Mairesse (1984), Pakes and Shankerman (1984a) and Mansfield (1984) provide recent examples where the focus on marginal productivity and the output elasticity of R&D, coupled with a general paucity of data, require the use of a more restrictive form.

5. A more general Cobb-Douglas form is

$$Q_t = Xe^{\theta t}L_t^{\alpha_1}K_t^{\alpha_2}Z_t \tag{3.2a}$$

where X is some constant. The most satisfying specification of this form for softwood plywood occurs when $X = 1$ and $\theta = 0$—which makes equation (3.2a) identical with equation (3.2). The form of equation (3.2) also has the advantage of expositional ease.

6. The inverted V lag structure is familiar in agriculture (and probably in field forestry). The diffusion of new research results is a longer process in agriculture than in the forest product industries. Extension agents share the information with many farmers (in contrast with the few firms in the SWPW industry which each make their own efforts to stay in contact with the Forest Products Laboratory). Initially a few risk-taking farmers try the new technology. If successful, it spreads to their neighbors in subsequent seasons and eventually reaches peak acceptance, before deteriorating.

7. Including disembodied technical change, as in equation (3.2a), adds the term $\gamma[\theta t - \lambda\theta(t - 1)]$.

8. For example, National Science Foundation (1981) data for selected years from 1957 through 1980 suggest an f approximating 0.006 for lumber, wood products and furniture. Subsequent estimation for SWPW (in chapter 4) suggests a λ around 0.9. In this case, the bracketed term is approximately 0.0006. In any case, this term is a constant.

9. The general case is:

$$\begin{aligned}
VMP_{rd,t}^{LR} &= (\mu P_t Q_t / E_t) \sum_{i=i_0+j_0}^{\infty} [\lambda^{i-i_0-j_0}/(1 + \rho)^i] \\
&= (\mu P_t Q_t / E_t)(1 + \rho)/(1 + \rho - \lambda)(1 + \rho)^{i_0+j_0} \\
&= \mu P_t Q_t /(1 + \rho - \lambda)(1 + \rho)^{i_0+j_0-1} E_t
\end{aligned}$$

The Softwood Plywood Industry

THIS CHAPTER applies the dual approach developed in the previous chapter to assess the impacts of public research investments in the softwood plywood (SWPW) industry. In particular, the objectives of this chapter are (1) to determine the total productivity of public research investments in softwood plywood, (2) to identify the distribution of research benefits and costs among consumers and producers, and (3) to estimate the value of the marginal product of research investments.

The answer to the first objective tells us whether the research investments were well-advised (*i.e.*, the value of increased output exceeds the research and development costs). The answer to the second tells us whether this research was well-advised as a public rather than a private investment (consumer gains are substantial but the public research costs exceed each producer's benefits). The answer to the third tells us whether there would have been net gains from an expanded research effort (the value of the marginal research product exceeds one).

When we began this analysis we also intended to assess the effects of the expansion of the SWPW industry into the South and to examine the distribution of benefits and costs among productive factors and among classes of consumers. Southern SWPW research is widely regarded as a great success while the more moderate success of SWPW research in general may be more comparable with the success of forest products research in general.[1] The more generalizable results from the national market should provide a broader reflection on the general social merits of public forestry research.

Our assessment of SWPW research begins with a background statement about the relevant features of the production process and the industry itself and proceeds with a restatement (from chapter 3) of the supply and demand functional forms from which the empirical estimation begins in this chapter. The central part of the chapter summarizes the data sources and econometric results. A final section converts these econometric results, together with our estimates of the research costs, into results responsive to the original three objectives.

We will find truly impressive results: net social benefits in the millions of dollars, rates of return possibly exceeding 300 percent and an initial return greater than $12 on the final public dollar invested. The first two results are unusually high in comparison with the general research evaluation literature. They encourage us to review the analysis in this chapter carefully. They also encourage us to compare these SWPW results with those for other forest products industries in order to find the norm for forest products research. (This will be a task for chapter 5.)

THE SOFTWOOD PLYWOOD INDUSTRY

Modern plywood mills are highly capital intensive. Workers operate automated machinery, although in many cases they can override, for example, a computer decision locating a cut on a sheet of veneer. Workers combine highly sophisticated machinery with minimal labor input to perform most productive operations.

Both private firms and the Forest Products Laboratory (FPL) of the US Forest Service, located in Madison, Wisconsin, conduct research designed to create technological improvements in the SWPW production process. The FPL has been a key actor in veneer and plywood research, especially from the mid-1940s through the mid-1960s, when the FPL pioneered research in SWPW production. It was only after the early 1970s that private firms engaged in research (such as the development of camera-computer interfaces for controlling lathe and veneer cutting operations and the generation of their own electricity from waste wood created during the production process) that was different from that conducted at the FPL.

In general, the relationship between FPL engineers and industry personnel has been close, with much communication regarding the orientation of applied research. Over the years FPL personnel have kept in touch with industry needs by regularly visiting plywood plants and by meeting with industry leaders at government installations.

FPL research is of an engineering nature. It tends to be applied research which is readily adaptable with only brief lags between research and implementation. Therefore, the adoption of new SWPW technologies is generally straightforward. Private firms stay in close contact with FPL personnel and actively seek information regarding research breakthroughs and the modification of production processes. Since FPL personnel release information about research in progress and since their research is completed rather quickly anyway, the lags between initiating new research, producing new knowledge and placing it in the hands of the industry are extremely short. The lag between new knowledge and implementation is, perhaps, almost instantaneous.[2] The average lags between completed research and full application are not more than one year for information dissemination and perhaps two years for application. Thus, we might expect FPL research to impact SWPW technology with a lag of two or three years.[3]

The structure of the SWPW industry itself is competitive and its focus is domestic. The largest firms in the softwood plywood and veneer industry (SIC 2436) share only 38 percent of the market (US Bureau Census 1978). This is less than the 40-50 percent usually considered indicative of market power. Furthermore, the product is homogeneous and the market is national in scope. Brands of plywood are undifferentiable on the basis of quality. (The American Plywood Association sets national standards for grading plywood and the various firms produce identical grades.)

Exports amounting to only 3.5 percent (5.5 million ft^2) of production in 1982, a normal year, support the contention of a domestic market. Table 4.1 provides further support. The twenty markets shown in this table represent 67.7 percent of all shipments in the United States for 1980. Many shipments were over considerable distance: from the western region to New York, Charlotte, Atlanta, Boston, and Miami; from the inland region to New York, Chicago, Philadelphia, and Detroit.

Moreover, moderate costs of entry into the industry also suggest a competitive market. Fleischer and Lutz (1962) claim that "[a] modern competing [softwood] plywood plant can be built for about $1½ million to make exterior-type sheathing-grade plywood," and Fleischer asserts that "[a] typical medium-size modern softwood plywood plant . . . will cost from $4 to $5 million."[4]

In sum, there is reasonable justification for believing that the SWPW industry is competitive. If the industry is competitive, then there is a conceptual basis for defining supply functions for firms within the industry and for calculating the marginal product of research inputs.

The major uses of plywood are for new housing construction, residential upkeep and improvements, and manufacturing. Plywood consumption increased rapidly in the 1950s and early 1960s, partly due to the substitution of SWPW for lumber in sheathing and subflooring for residential construction. Production peaked in 1972 at 18,324 million square feet (⅜" basis), reached a trough in 1974 at 15,878 million square feet, and more recently peaked again in 1978 at 19,964 million square feet (Anderson 1982).

The Douglas-fir region of Oregon and Washington was the major SWPW production region until 1963 when the industry began to expand into the South. The development of high speed lathes and lathe chargers made the utilization of smaller southern pine logs feasible. A large southern

Table 4.1. Top 20 markets for sanded, sheathing, and specialty softwood plywood, 1981

trading area	shipments M ft² 3/8" basis	% shipped by region		
		western	inland	southern
New York (4)	708,037	31.2%	15.3%	53.5%
Los Angeles (2)	676,470	83.3	16.7	--
Portland (1)	649,711	86.6	13.4	--
San Francisco/Oakland (3)	620,256	79.4	20.6	--
Charlotte (5)	594,967	14.2	0.2	85.6
Dallas/Ft. Worth (6)	572,340	13.3	0.9	85.8
Houston (8)	551,707	10.2	2.5	87.3
Export	538,675	26.2	21.6	52.2
Atlanta/Chattanooga (7)	456,158	9.8	0.9	89.3
Chicago (9)	443,804	22.1	13.4	64.5
Seattle (10)	428,131	76.4	23.6	--
Boston (14)	358,959	34.4	15.0	50.6
Philadelphia (11)	336,191	22.8	14.7	62.5
Minneapolis/St. Paul (16)	320,521	29.0	39.7	31.3
Miami (13)	320,336	18.9	0.7	80.4
Denver (15)	309,735	26.7	55.7	17.6
Detroit/Toledo (12)	300,897	19.2	15.0	65.8
Spokane (19)	289,790	13.4	86.2	0.4
Memphis (17)	277,415	8.7	2.1	89.2
Jacksonville (20)	270,493	11.4	1.1	87.5
total for 20 markets	9,024,593			
total U.S.	13,330,942			
share for top 20	67.7%			

Numbers in parentheses represent volume rank in 1980.
Source: Anderson (1982)

timber inventory, proximity to the large eastern markets, and lower labor costs all provided additional attraction for the industry (Holley 1969).

The first three southern operations began in the states of Arkansas and Texas in 1964. By 1967, there were 28 plants producing in Florida, Georgia, North Carolina, South Carolina, Virginia, Oklahoma, Arkansas, Louisiana, Mississippi, Alabama, Maryland, and Texas. The total number of SWPW plants in the United States increased during this period from 164 to 180, an increase of only sixteen, but the number of plants in the West declined from 161 to 152. By 1981, 66 plants, almost 38 percent of the US total of 176, were located in the South (Anderson 1982).

DATA AND ECONOMETRIC RESULTS

The previous section of this chapter provides the background that justifies our derivation of supply and demand and our choice of SWPW as a major benefactor of public research investments. This section restates the supply and demand functions derived in chapter 3 before reviewing the data and fundamental econometric results. It concludes with our estimates of public SWPW research costs.

The supply function, equation (3.7), with its (small and constant) bracketed term removed, is

$$q_t = (1 - \lambda)\ln A + \gamma(\alpha_1 + \alpha_2)(p_t - \lambda p_{t-1}) - \gamma\alpha_1(w_t - \lambda w_{t-1})$$
$$- \gamma\alpha_2(r_t - \lambda r_{t-1}) + \gamma\eta h_{t-2} + \gamma\mu g_{t-2} + \lambda q_{t-1} \qquad (4.1)$$

where we recall that A and γ are constant functions of α_1 and α_2. Their estimation proceeds accordingly. The demand function, equation (3.8), is

$$q_t = \beta_0 + \beta_1 p_t + \beta_2 w_{d,t} + \beta_3 r_{d,t} + \beta_4 p_{l,t} + \beta_5 p_{c,t} + \epsilon_{D,t} \qquad (4.2)$$

All exogenous variables are in log form. Their definitions and their data sources are:

p = price per thousand square feet, ⅜" basis (Ulrich 1983, Mills and Manthy 1974)

q = thousands of square feet on a ⅜" basis (Anderson 1982)

w = hourly wage; millwood, plywood, and structural members (Ulrich 1983)

w_d = hourly wage, residential building contractors (US Department of Labor 1979 and 1983)

r = user cost of capital, lumber and wood (Wharton Econometrics, personal correspondence)

r_d = Moody's Baa corporate bond rate (US Government Printing Office 1982)

p_l = price index for lumber (Ulrich 1983)

p_c = price per square foot of new residential construction (US Bureau of the Census 1975, 1983)

g = person-weeks devoted to softwood plywood research at the US Forest Product Laboratory (FPL) and regional Forest Service experiment stations (annual reports maintained by the FPL)

h = lagged total revenue, a proxy for private research investments

All data are annual aggregates for the United States from 1950 through 1980. Price and wage data are in real (1967) terms and interest rates are net of inflation, which is measured as the percentage change in the GNP deflator.

The two wage series, the price series for lumber and the R&D series all bear closer scrutiny.

Wages: Isolated missing observations in both wage series complicate the use of these series. The appendix reviews our estimation of the missing observations. In brief, regressing millwood, plywood, and structural member wages on a wage series for millwood, veneer, plywood and fabricated structural wood products generates missing estimates for *w*. Similarly, regressing residential building contractor wages w_d on all construction wages generates estimates for the missing observations in the w_d series.

Other productive factors: The demand function for SWPW derives from the supply side of the construction market. Therefore, it excludes variables from the demand side of the downstream (housing) market. For example, it does not include a real mortgage rate. The mortgage rate affects the demand for new construction. Therefore, its effect is embedded in the price of new construction p_c. If the mortgage rate as well as p_c were included in our regressions, then the two estimated coefficients (on p_c and the mortgage rate) would most surely be collinear and inefficient.

Those familiar with the construction industry will also note that while particleboard is important input in housing construction, there is no particleboard price series in the demand function. Ulrich (1983) provides producer price series for particleboard (both corestock and underlayment) but these series never perform well in our regressions—perhaps because particleboard became widely available commercially only late in our sample period. [See Bengston (1984) for a discussion.]

The effect of lumber price p_l is uncertain. Casual observation might suggest that lumber is a complement of softwood plywood in residential construction because increases in construction activity increase the demand for both goods. On the other hand, it may also be a substitute. The market for softwood plywood expanded rapidly in the 1950s and early 1960s as softwood plywood replaced lumber for sheathing and subflooring in residential construction (USDA Forest Service 1982, pp. 43-44). Therefore, we have no *a priori* hypothesis for the sign of the p_l coefficient.

R&D: Most studies of R&D productivity use R&D expenditures as a proxy for effort since the latter is usually unavailable. We are more fortunate except that the data for research person weeks *g* are in fiscal year form. FPL personnel involved in SWPW research projects from the 1950s through 1980 assisted in converting these data to a calendar year series.

Data on private R&D effort in the softwood plywood industry are unavailable. Lagged total revenue is our proxy for private R&D where the lag is less than or equal to the lag on more basic public research. Therefore, the proxy for private R&D, h_{t-2}, is determined endogenously as prior period total revenue (for reasons explained in chapter 3). (Appendix 3A reviewed the argument that firms conform to rule-of-thumb behavior in setting R&D expenditure levels at some share of cash flow.)

Preliminary estimates of the relevant public and private R&D lags are necessary before we proceed to estimate the full SWPW supply and demand system. John Lutz of the FPL suggests that the total lag is short.[5] Therefore, our preliminary estimates feature a possible range from one to seven years: from one to seven years for the lag on government research and from two to seven years for the lag on private implementation. That is, $i_0 = 2$, $3, \ldots, 7$ and $i_0 + j_0 = 2, 3, \ldots, 7$ in the formulation of equation (3.3).[6] There are 21 combinations. A linear, two-stage least-squares regression of equation (3.3) for each of the 21 combinations provides the preliminary test. The best fit in terms of correct signs and t-statistics occurs when both public and private lags are set at two years. This is perfectly consistent with Lutz' expectations. It is also consistent with our knowledge of the applied nature of SWPW research and it is compatible with Pakes and Schankerman's (1984) findings of various lags for applied R&D, all in the neighborhood of two years.

Reexamining the lag possibilities, subsequent to estimating the nonlinear supply and demand coefficients, leaves the preliminary estimates unchanged. The two-year lags still provide the best results.

General Results

Table 4.2 shows the nonlinear two-stage least-squares (NL2SLS) estimates for the demand and supply coefficients. The information in table 4.2 is all that is necessary to calculate the value of the marginal product (*VMP*) for labor and capital. This information and our public R&D series are all that is necessary to calculate measures of the returns to public SWPW research. The remaining discussion of the table 4.2 results only provides background and confidence for our final results.

The imposition of theoretical prior cross constraints on the supply coefficients mitigates collinearity problems in the NL2SLS system. One-tailed t-tests are appropriate for the supply equation/production function because the anticipated signs on all coefficients are positive.

The first set of results reflects the more general Cobb-Douglas form described by equation (3.2a) where θ, the coefficient on the time variable, reflects disembodied technical change. This coefficient is statistically

Table 4.2. NL2SLS estimates of demand and supply coefficients for softwood plywood

results with $\theta \neq 0$		results with $\theta = 0$	
demand	supply/production	demand	supply/production
(β_0) 8.7981** (3.8658)	labor (α_1) .2199** (.1183)	(β_0) 8.8088** (3.8696)	labor (α_1) .2057** (.1125)
P_t (β_1) -2.7016* (.4895)	capital (α_2) .1072*** (.0758)	P_t (β_1) -2.7034* (.4901)	capital (α_2) .1263** (.0604)
$w_{d,t}$ (β_2) -0.6072*** (.4203)	H_{t-2} (η) .0550*** (.0346)	$w_{d,t}$ (β_2) -0.6076*** (.4205)	H_{t-2} (η) .0660* (.0218)
$r_{d,t}$ (β_3) .1262** (.0597)	G_{t-2} (μ) .0213*** (.0159)	$r_{d,t}$ (β_3) .1263** (.0597)	G_{t-2} (μ) .0247** (.0131)
$P_{l,t}$ (β_4) 3.0589* (.4260)	lag (λ) .8964* (.0779)	$P_{l,t}$ (β_4) 3.0603* (.4265)	lag (λ) .8694* (.0378)
$P_{c,t}$ (β_5) 2.3618*** (1.5572)	time (θ) -.0128**** (.0403)	$P_{c,t}$ (β_5) 2.3583*** (1.5586)	
R^2 .8874	.9841	.8869	.9839
Durbin-Watson 1.46	---	1.46	---
Durbin's h --	-.8432	--	.0270
degrees of freedom 22	22	22	23

Numbers in parentheses are standard errors.
* Significant at the 1 percent level.
** Significant at the 5 percent level.
*** Significant at the 10 percent level.
**** Insignificant and incorrect sign.

insignificant. Therefore, there are few unaccounted technical improvements and disembodied technical change is unimportant in the SWPW case.[7]

We can accept the simpler production form of equation (3.2) and the basic form of equation (4.1) as initially given in this chapter. In this case $\theta = 0$ and the right hand columns of table 4.2 show the appropriate results. The coefficients are not greatly changed and the regression results are stronger. All coefficients in the supply equation have the expected signs and all are significant at the 5 percent level or better. The small value of Durbin's h statistic indicates that serial correlation is not a problem in the supply equation.

The estimated output elasticity for government R&D (μ) is the most important supply coefficient for our purposes. Its standard error declines when we constrain θ to equal zero, but it remains large relative to the estimated standard errors for other inputs. This encourages us to examine subsequent analytical applications of this result for sensitivity by considering cases where the output elasticity of government R&D is one standard deviation lower than the actual estimated coefficient.

The interpretation of the regression results for the demand equation is straightforward. The coefficients for all but two variables are significant at the two-tailed 1 percent probability level. These two are the coefficients for construction wages β_2 and the price of finished goods β_5. They may exhibit multicollinearity since wages comprise a large portion of housing prices—which comprise the largest share of finished goods. Nevertheless, theory supports the strong price expectations (1) that construction labor is a complement to SWPW in consumption, therefore that β_2 is negative and (2) that β_5 is positive. Thus, one-tailed tests are appropriate. Both coefficients are significant at the one-tailed 10 percent level. The Durbin-Watson test for serial correlation in the demand equation falls in the inconclusive range at a 5 percent level of confidence. Therefore, correction for serial correlation is unnecessary.

The coefficient of the real interest rate β_3 is positive. This variable is a composite short- and long-term rate. It reflects the opportunity cost of capital inputs (such as construction materials and machinery) in construction. The positive sign on β_3 is consistent with the expectation that the effect of SWPW substitutes in demand; such as particleboard, insulation board, and hardboard; combine to override the effect of complements in production (such as plant and equipment). Of course, the plant and equipment costs for residential construction are probably small. The coefficient of the real interest rate is small, suggesting nearly offsetting effects.

The form of the demand equation does not restrict the sign of the lumber price coefficient β_4. The regression results suggest that lumber is a net substitute for softwood plywood. This observation agrees with the majority of previous econometric evidence on cross-substitution between lumber and plywood (*e.g.*, McKillop, Stuart, and Geissler 1980). As discussed earlier, softwood plywood replaced lumber in sheathing subflooring when softwood plywood became relatively cheaper. Lumber remains a substitute since, if the relative price of lumber were to fall, it can replace softwood plywood in these uses.

Output Elasticities of Inputs and Supply and Demand Elasticities

This section investigates the output elasticities for the various productive inputs as well as the price elasticities for supply and demand. The coefficients of labor α_1 and capital α_2 in table 4.2 are the respective elasticity estimates since the output elasticity of an input under a Cobb-Douglas technology is simply the associated exponent.

The output elasticity for government research is the infinite summation

$$\sum_{i=0}^{\infty} \lambda^i \mu = \mu/(1 - \lambda)$$

The magnitude of the coefficient for the research lag ($\lambda = 0.8694$ in table 4.2) indicates that research inputs endure for a long time. The standard error of a nonlinear function $f(x)$ is

$$[(\partial f/\partial x)\Gamma(\partial f/\partial x)']^{1/2}$$

evaluated at the estimated x. In this case, $x = (\mu,\lambda)$ and Γ is the 2×2 variance matrix associated with x. In sum, the output elasticity for government research is 0.19. Its standard error of 0.13 (with a t value of 1.46) indicates significance at the 10 percent level in a one-tailed test.

The output elasticity for private R&D cannot be computed from the regression results because the regressions use a proxy for private R&D. Nevertheless, an upper bound for this elasticity can be estimated as a residual. Since the summation of all output elasticities cannot exceed unity, the output elasticity of private R&D must be less than 0.48.[8] If the output elasticity of private R&D in the SWPW case were 0.24 (the midpoint of the estimated range), then the coefficient of returns to scale would be 0.76.

These estimates are comparable to other estimates in the research evaluation literature. For example, Griliches and Mairesse (1984) use the production function approach to estimate output elasticities for private R&D in various industry aggregates. They often find sharply diminishing returns to scale—in one case the summation of output elasticities of capital, labor, and R&D is only 0.69—and they estimate output elasticities of private R&D which range from around 0.03 to nearly 0.30.

The regression results can also provide the price elasticities of supply and demand. The supply price elasticity is

$$(\alpha_1 + \alpha_2) / (1 - \alpha_1 - \alpha_2)$$

Its estimated value is 0.497. The standard error of the estimate is 0.152 (with a t value of 3.270). The demand price elasticity can be read directly from table 4.2 as $\beta_1 = -2.703$.

The demand price elasticity for softwood plywood is greater than previous estimates for composite (hardwood and softwood) plywood in general. McKillop, Stuart, and Geissler (1980) estimate an elasticity of -0.667 while Rockel and Buongiorno (1982) estimate an elasticity of -0.95. These composite plywood elasticities are not strictly comparable with our SWPW estimate because SWPW and hardboard plywood are not close substitutes. SWPW is used mainly for sheathing and subflooring while hardwood plywood is used extensively for paneling and furniture manufacture. Nevertheless, if they were close substitutes, then our results would be expected, since the less aggregate demand curve should be more elastic.

The Cost of Public Research

We are fortunate to have annual reports from the FPL which record government and cooperating non-government scientist time spent on SWPW research and publication for the entire period from 1950 to 1980. Multiplying scientist time by the cost of that time provides an estimate of direct outlays for research effort. The additional costs for private firms to receive, review, and apply the results of government research are unavailable. (These particular private costs are distinct from the costs of that private sector R&D which private firms initiate independently.) We construct a multiplier, representing the additional costs of implementing public R&D, in order to incorporate these induced private sector costs.

There are four steps to transforming SWPW research into a factor of production: the research itself, dissemination of research findings, evaluation by the private sector, and the technical engineering adjustments necessary for implementation. FPL annual reports record the first and the public share of the second. Our multiplier reflects the private share of the second plus all expenditures on the third and fourth steps. Dissemination costs may be great for agricultural markets or field forestry because there are many private farmers or forest managers to inform. *Dissemination* is less expensive for SWPW and for the forest products industries in general. The number of SWPW firms is small and most firms maintain close contact with the FPL. On the other hand, SWPW *implementation* costs may be greater than implementation costs in agriculture or field forestry because the former may impose larger expenditures on capital equipment.

The determination of total annual R&D costs E_t is

$$C_t = (1 + mN_t)C_t'$$

and $E_t = C_t G_t$ \hfill (4.3)

where G is the scientist weeks of research, C' is the public cost and C is the full cost per scientist week, N is the number of plants (not firms) in operation, and m is the implementation cost per plant for each government research dollar. N is available in Anderson (1982). C' can be constructed from an academic R&D price index (Sonka and Padberg 1979) and Callaham's (1981) estimated 1977 cost per US Forest Service scientist year (the US Forest Service funds the FPL) plus overhead cost estimates supplied by the FPL. It must be deflated to 1967 dollars by the all-commodities price index in order to maintain consistency with other prices in this analysis.

General evidence about m is unavailable. Therefore, our choice of m refers to a particular case, auxiliary torque, which has the advantage of known research implementation costs. (We interviewed two FPL personnel and ten industry specialists who all believe it to be representative.) The estimated m is 0.26 which means that private firms spend 26 cents per plant on dissemination, evaluation and implementation for every dollar of public research.[9] Multipliers of 0.13 and 0.39 test for the sensitivity of our final results to this estimate.

Table 4.3 shows the cost series developed with these assumptions. These estimates are the sum of direct government research costs and private implementation costs. They appear as C_iG_t in equations (3.15) and (4.3). The table shows that SWPW research expenditures increased until the late 1960s when research emphasis at the FPL shifted to particleboard. SWPW research expenditures did not recover to their previous levels until the late 1970s.

THE RETURNS TO SOFTWOOD PLYWOOD RESEARCH

Table 4.2 shows the supply and demand functions that permit estimates of the gross benefit flows from public SWPW research. Table 4.3 provides the R&D cost estimates necessary, in addition to the table 4.2 results, to derive our final results. These results are in terms of the value of the marginal product [VMP, equations (3.10) and (3.12)], the marginal internal rate of return [$MIRR$, equation (3.14)] and the net economic benefit [NPV^{NEB}, equation (3.26)].

Figure 4.1 displays annual values for the initial impact, or short-run VMP as derived with an undiscounted form of equation (3.10). The shift away from SWPW research in the late 1960s and rising real prices and quantities of plywood explain the increase in its marginal productivity. More importantly, this increase is also a result of the concave nature of the Cobb-Douglas production function. Decreasing the level of research inputs raises the marginal productivity of the remaining research inputs.

Table 4.4 shows the geometric means of the annual short-run (and long-run) VMP series. We can compare these short-run SWPW results with other results from the agriculture literature. Griliches (1964) estimates an average VMP for aggregate agriculture research expenditure of $13 in 1949 dollars or $16.52 in 1967 dollars. Peterson (1967) reports an average VMP for poultry research of $18.52 in 1955 dollars or $21.09 in 1967 dollars. Griliches' and Peterson's results fall in the range of $8.40 to $24.33 for our average values for public SWPW research.

Figure 4.2 displays annual results for the long-run *VMP* for SWPW

Table 4.3. Total cost estimates for public softwood plywood research (in 1967$)

year	multiplier		
	.13	.26	.39
1950	455,763.01	865,208.64	1,274,654.28
1951	462,491.00	882,975.56	1,303,460.12
1952	324,327.03	662,307.43	920,287.83
1953	371,447.53	714,797.67	1,058,147.80
1954	412,966.00	796,703.89	1,180,442.78
1955	569,313.87	1,102,039.45	1,634,765.03
1956	750,256.02	1,456,012.88	2,161,769.73
1957	825,455.49	1,600,792.24	2,376,129.00
1958	838,958.76	1,630,357.51	2,421,756.25
1959	1,015,149.36	1,978,132.78	2,941,116.19
1960	862,111.15	1,682,694.79	2,503,278.42
1961	863,891.05	1,685,906.63	2,507,922.21
1962	2,171,838.40	4,240,989.41	6,310,140.42
1963	1,978,659.69	3,864,901.84	5,751,143.99
1964	2,300,368.92	4,497,674.72	6,694,980.52
1965	1,753,452.35	3,432,688.79	5,111,885.23
1966	1,058,532.78	2,074,588.33	3,090,643.87
1967	1,350,988.50	2,646,608.62	3,942,228.74
1968	827,337.21	1,619,839.17	2,412,341.13
1969	698,040.91	1,367,008.89	2,035,976.87
1970	582,952.78	1,141,886.07	1,700,819.37
1971	133,788.24	262,344.25	390,900.26
1972	273.830.92	537,113.65	800,396.38
1973	480,844.10	943,165.70	1,405,487.30
1974	1,108,275.90	2,174,492.00	3,240,708.10
1975	950,163.78	1,863,542.28	2,776,920.79
1976	903,040.76	1,769,653.91	2,636,267.05
1977	471,928.85	925,018.22	1,378,107.60
1978	1,661,916.02	3,256,438.84	4,850,961.67
1979	1,604,074.33	3,144,771.57	4,685,468.80
1980	915,400.68	1,793,284.94	2,671,169.20

Source: FPL attainment reports

Table 4.4. Geometric mean value of marginal products (short-run and long-run) and marginal internal rates of return for publicly funded softwood plywood research

cost multipliers	VMP^{SR}	VMP^{LR}			MIRR
		at social discount rates of			
		.04	.07	.10	
.13	24.33	137.14	113.36	95.92	438%
.26	12.49	70.38	58.17	49.23	299%
.39	8.40	47.33	39.13	33.11	236%

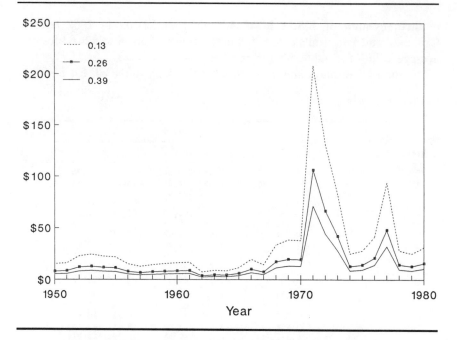

Figure 4.1. Short-run value of the marginal product for public SWPW research, with varying private adoption cost multipliers

research with an intermediate multiplier of 0.26. These values are calculated according to equation (3.12) using social discount rates of 4, 7, and 10 percent, the rates preferred for project evaluation by the US Forest Service, the Water Resources Council and the Office of Management and Budget, respectively. The long-run average *VMP*s are much higher than the short-run *VMP*s (see table 4.4) because the long-run *VMP*s include returns accumulating for several years subsequent to the initial research payoff. Long-run *VMP*s range from 3.9 times higher than the short-run *VMP* for a 10 percent discount rate to 5.6 times higher for a 4 percent discount rate. The average *VMP* is sensitive to the assumed multiplier. Doubling the multiplier from 0.13 to 0.26 decreases the average *VMP* by about half. Increasing the multiplier to 0.39 decreases the average *VMP* by approximately one-third more.

The *MIRR*, on the other hand, is relatively less sensitive to changes in the multiplier. Its estimates, also shown in table 4.4, range from 236 percent for a multiplier of 0.39 to 438 percent for a multiplier of 0.13. Doubling the multiplier from 0.13 to 0.26 only decreases the *MIRR* by approximately one-third. The smallest computed *MIRR* (236 percent) exceeds any reported *MIRR* in the agriculture literature. For example, Otto and Havlicek (undated) report *MIRR*s of 152 to 210 percent for corn research and 148 percent for wheat. Otto (1981) reports *MIRR*s of 101 percent for sorghum and 176 percent for soybeans. Davis (1981) reports an *MIRR* of up to 154 percent for aggregate agricultural research.

Table 4.5 shows the net economic returns in 1967 dollars and table 4.6 shows the *IRR* associated with these returns. These results, in all cases, reflect the coefficient of public R&D estimated in the constrained regression ($\mu = 0.0247$). Results for a one standard deviation reduction in that coefficient ($\mu = 0.0116$) are also shown for comparison. The social discount rates of 4, 7, and 10 percent and the range of multipliers; 0.13, 0.26, and 0.39; continue as before. Internal rates of return in table 4.6 reflect both consumers' surplus and the sum of consumers' and producers' surpluses. The *IRR*s have unique solutions since the time streams of net benefits change signs only once.

The estimated returns to research are large compared with results from the many studies of agricultural markets where internal rates of return between 90 and 100 percent are on the high side [but not uncommon

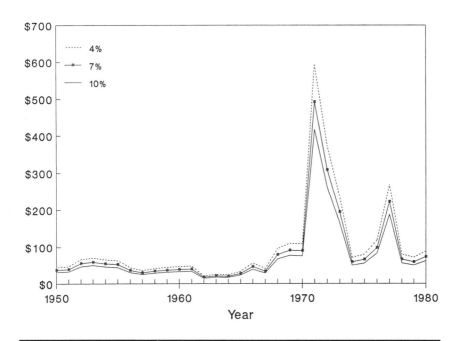

Figure 4.2. Long-run value of the marginal product for public SWPW research, cost multiplier = 0.26 with varying social discount rates

(Ruttan 1980)]. Our higher estimates may be explained by institutional differences. It turns out that they are not a result of our methodological improvements.

SWPW research over our time period tends to be a small scale operation, entirely conducted by a very few individuals. It is also applied, rather than basic, in nature. Therefore, firms are able to adopt the new techniques quickly. In addition, firms take the initiative in staying abreast of the new developments. In contrast, agricultural research is often more basic and may require years of research before its eventual application. The need to inform many farmers who do not themselves maintain active ties with the research stations introduces another lag. Combining the various

Table 4.5. Net returns to public research investment in softwood plywood research, 1950-1980 (in millions of 1967$)

multiplier		$\mu = .0247$ social discount rate			$\mu = .0116$ social discount rate		
		.04	.07	.10	.04	.07	.10
.13	NPV^{CS}	2,428.2	1,329.0	788.0	1,119.7	611.5	361.8
	NPV^{PS}	2,765.1	1,513.7	897.7	1,276.3	697.3	412.7
	NPV^{NEB}	5,208.8	2,853.2	1,693.1	2,411.5	1,319.3	781.9
.26	NPV^{CS}	2,413.4	1,319.1	781.0	1,105.0	601.6	354.8
	NPV^{PS}	2,750.4	1,503.8	890.7	1,261.5	687.4	405.7
	NPV^{NEB}	5,194.0	2,843.3	1,686.1	2,396.8	1,309.4	774.9
.39	NPV^{CS}	2,398.7	1,309.2	774.0	1,090.2	591.7	347.7
	NPV^{PS}	2,735.6	1,493.9	883.6	1,246.8	677.5	398.6
	NPV^{NEB}	5,179.3	2,833.4	1,679.1	2,382.0	1,299.5	767.9

$NPV^{CS} + NPV^{PS} + E = NPV^{NEB}$ since each of the net present values on the left-hand side are calculated by subtracting total cost.

Table 4.6. Internal rates of return for investments in softwood plywood research, 1950-1980

multiplier	estimated returns when $\mu = .0247$		estimated returns when $\mu = .0116$	
	IRR^{CS}	IRR^{NEB}	IRR^{CS}	IRR^{NEB}
.13	468%	707%	302%	464%
.26	326%	499%	207%	323%
.39	260%	402%	163%	258%

lags suggests heavy discounting for some agricultural research benefits. We expect the returns for agricultural research to be lower than the returns to softwood plywood research. Indeed, the returns to forest products research activities in general may range on the high side of returns to agricultural research for this reason.

We can inquire whether our approach, which estimates all relevant elasticities endogenously and which re-estimates productive shifts for each year in the research period, is also responsible for our higher estimates of research benefits. Previous analyses generally use observed prices and quantities rather than actually solving for the average "push" imparted by research upon the supply curve. Strict comparability between our results and SWPW results using the previous Bengston-Griliches (B-G) approach (see chapter 3) is difficult. Nevertheless, we can approximate the B-G results for SWPW using our demand price elasticity, assuming an infinite supply price elasticity, our stream of costs, and the alternate assumptions that benefits last (a) one period or (b) indefinitely. The B-G net present values for both periods are greater than our net present values for both the consumers' surplus and the net economic benefit measures for 17 of 29 individual years and for the 1950-1980 period as a whole. This lends further confidence that our own SWPW estimates, while large, are not unreasonable.

This comparison also suggests that the observed SWPW price (and the actual supply curve) is lower after the innovation due to decreases in real factor prices—as well as due to research breakthroughs. In other words, previous approaches would credit research with some productivity gains actually due to falling real factor prices. Wage adjustments in the SWPW industry in general, and particularly the decreasing wage as the industry moved from the Northwest to the South, reinforce this point. The previous approach would not reflect these wage decreases but would erroneously credit their impacts to research and, as a result, would show larger final returns to SWPW research.

SUMMARY AND CONCLUSIONS

This chapter applies the method developed in chapter 3 to assess the benefits of public research investments in the softwood plywood industry. It begins by reviewing the development of the industry over the last 35 years, including its tremendous growth and its geographic shift from the Pacific Northwest to the South, and by assessing its general competitiveness.

The analysis itself begins by outlining the data needs and data modifications necessary to simultaneously estimate demand and the supply-production dual. The key results from this estimation are:

supply price elasticity	0.497
demand price elasticity	−2.703
labor output elasticity	0.2057
capital output elasticity	0.1263
public research output elasticity	0.0247 (initial)
	0.19 (long-run)
private research output elasticity	<0.48

The first six elasticities are all significant at the 10 percent level or better. The private R&D output elasticity is a residual. The residual estimate of private R&D elasticity declines if we anticipate larger scale effects in production. R&D has an initial effect in the period immediately following its implementation and an accumulating longer-term effect before the research breakthrough deteriorates over time. The initial and the long-run public research elasticities reflect this distinction.

This information, together with information on research costs, provides the necessary background for determining all the standard measures of returns to public research. A separate section of the chapter reviews our estimate of research costs. This estimate relies on both documentation of the public research effort by the US Forest Service's Forest Products Laboratory and a more casual approximation of the level of private expenditures necessary to implement public research breakthroughs in the mills. The mean values (private implementation multiplier = 0.26) for our estimates are:

VMP (short-run)	$12.49
VMP (long-run)	$58.17 (7% discounting)
MIRR	299%
NPV: consumers surplus	$1.319 (billion 1967$ at 7%)
NPV: producers surplus	$1.503 (billion 1967$ at 7%)
NPV: net economic benefit	$2.843 (billion 1967$ at 7%)
IRR on *NPV:* consumers surplus	326%
IRR on *NPV: NEB*	499%

The value of the marginal product (*VMP*) measures the incremental gain in output due to the last public research dollar spent. The short-run *VMP* is comparable to those reported for agriculture research. It shows that there was considerable opportunity for expanding SWPW research investments, as additional research investments would still yield more than their investment costs.

The long-run *VMP* and the marginal internal rate of return (*MIRR*) are much greater than generally found in agriculture. They provide further support for the argument that there would have been substantial social gains resulting from additional investments in SWPW research. These gains may be larger than those reported in agriculture for two reasons; shorter lags between research and implementation, thus less discounting of benefits, and the smaller scale of research investment attached to significant SWPW research breakthroughs and productivity gains. (The previous research evaluation technique used in the agriculture literature would actually estimate even greater new research benefits in the SWPW case—perhaps because it would not fully account for wage decreases as the industry moved South.)

The net present value (*NPV*) calculations further display the success of public investments in SWPW research. They also show that consumers and producers shared almost equally in the benefits of this research.

The *NPV* for producers' surplus of $1.503 billion (for a social discount rate of 0.07 and a research cost multiplier of 0.26) suggests that, in an industry with approximately eighty firms, the average firm gained approximately $18.8 million over the thirty-year period 1950-1980, or approximately $0.6 million annually. This annual gain is insufficient to cover the full R&D costs in any of the first eighteen years of our analysis, or before 1968. Thus, the average firm could not have justified the R&D costs by itself—unless it had an unusually long time horizon—and neither the firm nor society would have received the large net gains from SWPW research had this research been left to a competitive private sector. In sum, public research effort was clearly productive in the SWPW case and society, including both consumers and producers, benefitted.

This analysis provides an historical view of an industry in a period of rapid growth. Public research investments contributed a major share (0.19) to this growth experience and we are confident that larger public research investments would have been even more rewarding.

These observations do not, however, carry over as arguments for expanded future public research in SWPW. That is another case. It is also a less likely case. There were few cost-effective substitutes for SWPW during most of the period of our analysis, but structural particleboard (SPB) has become an important substitute and the SPB industry has become a large industry (relative to the SWPW industry). Future returns to SWPW are likely to be lower because the output resulting from future SWPW research will only substitute for SPB output. It will not add substantially to social welfare. Nevertheless, even somewhat lower returns might still be in excess of what we generally observe for either public or private investments.

APPENDIX: WAGE DATA

There are three different wage series for SWPW consumers: average hourly earnings for construction workers, for general building contractors, and for residential building contractors (US Department of Labor 1979, 1983). The latter two series are incomplete for the period 1950-1980. In order to estimate the missing observations, we fit ordinary least squares (OLS) regressions of the two incomplete wage series on construction worker wages both with and without a time index. Where there is evidence of serial correlation, we estimate a first order autoregressive process. Table 4A.1 records the results of the most satisfying equations. The residential building contractor wage series with missing data generated from general building contractor wages (which, in turn, have missing data generated from construction worker wages) provides the best fit for our SWPW supply and demand system.

Construction of the wage series for SWPW workers is similar to the construction of a wage series for residential building contractors. First, we regress a series of the average hourly wages for millwood, veneer, plywood, and fabricated structural wood products (MVPFSWP) (Phelps 1977) on the wage series for plywood and veneer (PV), both with and without a time index, using OLS and a first order autoregressive process. The OLS regression performs well and can be used to complete the MVPFSWP series for 1950 through 1980. We then regress a wage series for millwood, plywood and structural members (MPSM) (Ulrich 1983) on the completed

Table 4A.1. Regression estimates for constructing missing data on building contractor wages

	general building contractor		residential building contractor
intercept	.1723 (.0629)	intercept	.4089 (.1161)
construction workers wage	.9076 (.0101)	general building contractors wage	.9012 (.0159)
lag coefficient	-.7329		
R^2	.9975	R^2	.9978
		Durbin-Watson	.8676
degrees of freedom	20	degrees of freedom	7

Numbers in parentheses are standard errors. The series for general building contractor wages was fitted by a first order autoregressive process, while the series for residential building contractor wages was fitted with OLS. A first order autoregressive process was estimated in the latter case, but it changed the coefficient only slightly while the *t* statistic for the autoregressive term was not very significant at -1.39.

MVPFSWP series in the same manner and complete the MPSM series for 1950-1980. Table 4A.2 records the best estimates. The SWPW supply equation uses the revised MPSM series.

Table 4A.2. Regression estimates for constructing missing SWPW wage data

	MVPFSWP		MPSM
intercept	.1515 (.0092)	intercept	-.7218 (.1288)
PV	.9784 (.0096)	MVPFSWP	1.3214 (.0256)
time	.0032		
R^2	.9988	R^2	.997
Durbin-Watson	2.0618	Durbin-Watson	1.4031
degrees of freedom	16	degrees of freedom	7

Numbers in parentheses are standard errors. Both series were estimated with OLS procedures.

NOTES

1. Both conceptual and econometric problems frustrate the examination of the southern industry. Conceptually, it is difficult to distinguish the market for southern pine plywood from the market for all other softwood plywood. Indeed, consumers do not usually distinguish qualitatively between softwood plywood produced in different regions. Rather, consumers widely substitute southern and western pine plywood in construction, the primary use of plywood from either region. Furthermore, econometric attempts to distinguish southern pine plywood research from all other softwood plywood research yield incorrect (but insignificant) signs on several supply coefficients, notably the critical own price and research coefficients. Finally, the lack of cross-sectional data frustrates distributive distinctions among the various intermediate and final consumers of softwood plywood.

2. H.O. Fleischer, personal interview, March 3, 1983.

3. J.F. Lutz, personal interview, March 3, 1983.

4. There is some contradiction of the contention that the industry is competitive. It comes in the form of a successful 1978 conviction of three plywood producers for conspiracy to fix prices. The industry surely is noncompetitive if as few as three producers can successfully fix prices for the entire industry.

This conviction is probably not justified. Haddock (1982) argues that conviction was the result of a poor understanding of base point pricing, rather than evidence of market power. Base point pricing refers to the practice of quoting prices from a common origin, or "base point," for the industry regardless of the location of the firm making the sale. The base point provides a uniform reference for buyers and, therefore, is a means for producers to insure product comparability. Base point pricing for freight charges can easily arise in competitive industries.

5. Personal interview, March 3, 1983.

6. One year public research lags create perfect collinearity when lagged price and quantity enter the Koyck transformation.

7. Disembodied technical change introduces an additional term, $\gamma[\theta t - \theta(t - 1)]$, in the estimation of equation (4.1). It also leads to a potentially confusing result in the SWPW case. The estimated θ is small and negative—suggesting negative, unaccounted for, technical improvements—but it is also statistically insignificant.

The poor performance of θ is initially surprising because we expect it to account for quality changes in the non-research inputs like raw materials and capital equipment. We might also expect that R&D having to do with these items is funded by private industrial sources. These expectations would be unfounded. Government R&D is extensive over this period. It includes efforts to lower raw material costs by experimenting with little used logs, new gluing techniques, and virtually all successful new capital equipment experiments except camera-computer interfaces. Even the camera-computer developments were funded in Sweden—outside the boundaries of our domestic US analytical model. Therefore, the R&D index accounts for virtually all of the technical change.

8. $0.48 = 1 - 0.2057 - 0.1263 - 0.19$ (the output elasticities of labor, capital and government research, respectively).

9. Seldon (1985) provides detail.

Further Examinations in Forest Products Industries

THIS CHAPTER repeats the analysis of chapters 3 and 4 for three additional forest product industries; sawmills, woodpulp, and wood preservatives. The sawmill and woodpulp industries are interesting as the two major consumers of wood and wood fiber. These two industries largely define the forest products sector of the US economy. Therefore, successful research in these two industries should have a telling impact on the forest products sector as a whole. The sawmill industry is a model of competitive activity while the pulp industry is a more concentrated industry. It will be interesting to observe whether this contrast in competitiveness has implications for either the level or the efficiency of research investments.

The wood preservatives industry is a much smaller component of the forest products sector than either the sawmill or woodpulp industries. Its interest for us is in its representation of an important special case that is often repeated in the sector. Large shares of all wood preservatives research are product-altering research. They do not decrease the production costs of an existing product. Therefore, our benefit measures are not well-designed to capture full estimates of these research shares and our final estimates of aggregate wood preservatives research will be most conservative. Because most research in the full forest products sector of the economy, unlike wood preservatives research, is *cost-reducing rather than product-altering*, then the wood preservatives research benefit estimates may provide a lower bound for expected benefit estimates across the entire sector.

This chapter moves directly to the empirical tests for each indus-try—with additional comment only where the data require variation from the technical approaches in chapters 3 and 4. The period of analysis, 1950-1980, remains the same for all three industries. A final section of the chapter reviews our results and inquires into possible reasons for any distinctions between industries.

SAWMILL RESEARCH

Sawmills were among the original processors of industrial timber. They have been important in this country since the earliest colonial days. The sawmill industry has always been competitive, with relatively low entry costs and many mills, each located close to its basic timber resource.

The value of sawmill output increased only 18 percent over the thirty year period of our inquiry [from $4.0 billion (1967$) in 1950 to $4.7 billion in 1980]. The construction industry consistently consumes three-fourths of annual sawmill output and home improvement consumes much of the remainder. The four largest firms' share of production rose from under 11 percent in 1963 to more than 17 percent in 1978 while the number of mills declined by two-thirds and the number of millworkers declined by one-third between 1950 and 1983 (US Department of Commerce 1985a). Most of this adjustment reflects increases in mill capacity and the closing of many smaller, less efficient mills between the late 1950s and the middle 1970s. In sum, mill scale has increased but the industry remains highly competitive—with relatively low initial capital costs and low restarting costs, easy entry and frequent exit and re-entry of marginal firms following signals from the construction industry and the aggregate economy.

These adjustments in industry structure reflect a period of research and technical change concentrating on labor costs and labor productivity. Technical change in this period initially featured mechanical debarking and gravity feeds of logs along the saw carriage. These techniques made water storage of log inventories unnecessary and decreased manual handling of the primary material. The subsequent introduction of chip 'n saws, narrower saw blades and tungsten-tipped blades all improved wood utilization rates.

The high technology electrical and computer revolution in the sawmill industry began in the late 1950s. Electromechanical stress graders now permit more precise log grading before sawing. Mechanical carriages and kickers permit sawyers to manipulate and position logs more easily. Computer scanning and sorting techniques, introduced in the late 1960s and early 1970s, improve measures of output quality. The best-open-face (BOF) technology now used in 50-60 percent of all mills employs a computer to determine the configuration of saw slices that maximizes lumber yield from each log. This improvement alone increased physical yield 10 percent (Horvath 1980).[1] The computer also controls log movement on the saw carriage without manual assistance.

The net effect of these various innovations (aside from the important increase in scale of operations) has been to reduce labor costs sharply. There is now very little manual handling of the log from the time it enters the debarker, through the sawing process and to the sorting of the lumber output. Absolute labor inputs fell for the industry as a whole for all but six years between 1950 and 1980 and the average skill level of remaining millworkers is much greater now than it was in 1950.[2]

Supply and Demand

The supply function, comparable to equation (3.7) in chapter 3 with its (small and constant) bracketed term removed, is

$$
\begin{aligned}
q_t = (1 - \lambda)\ln A &+ \gamma(\alpha_1 + \alpha_2)(p_t - \lambda p_{t-1}) - \gamma\alpha_1(w_t - \lambda w_{t-1}) \\
&- \gamma\alpha_2(r_t - \lambda r_{t-1}) + \gamma\eta h_{t-i} + \gamma\mu g_{t-5} + \lambda q_{t-1}
\end{aligned} \tag{5.1}
$$

where γ and A are constant functions of α_1 and α_2. All exogenous variables are in logarithmic form. Their definitions and their data sources are

q = sawmill and planing mill (SIC 2421) output (provided by E. Henneberger, Bureau of Labor Statistics)

p = a price index for SIC 2421: value of shipments divided by output (US Department of Commerce 1985a), deflated by the producers' price index

w = average hourly wage for SIC 2421 production workers (US Department of Labor 1979, 1983), deflated by the producers' price index

r = real user cost of capital for structures in lumber and wood products (SIC 24) (Wharton Econometrics, personal correspondence)

h = a proxy for private sawmill R&D expenditures: real total revenue, lagged appropriately

g = government scientist months for sawmill research (various FPL attainment reports)

These exogenous variables each require further discussion. The SIC 2421 definition changed in 1958. Therefore, satisfactory quantity and price measures and wage series only exist for 1958-1980 and we must reconstruct observations for 1950-1957. We projected the price index backward from 1958 according to adjustments in the producer price index for lumber (Ulrich 1983). Ulrich's series for lumber includes the products of SIC 2421. Adjusting the 1950-1957 values of shipments by the ratio of the new to the old definition values for 1958, and dividing by the price proxies for 1950-1957 generates the necessary backward extension for the quantity measure.

The missing wage observations can be estimated from wages in SIC 242 (US Department of Labor 1979, 1983) and time using the regression

$$\log w_t \text{ (SIC 2421)} = -20.97 + 1.01 \log w_t \text{ (SIC 242)} + 2.77 \log \text{year}$$
$$\phantom{\log w_t \text{ (SIC 2421)} = }(3.94)\quad(0.01)\phantom{\log w_t \text{ (SIC 242)} + }(0.52)$$

The numbers in parentheses are standard errors. All coefficients are significant at the 1 percent level. The coefficient of determination is 0.9996. The Durbin-Watson statistic (1.492) is in the indeterminate range, reflecting uncertainty with respect to serial correlation.

The user cost of capital term is from a new Wharton series which separates the SIC 24 cost of capital into its equipment and structures components. A summary series was the preferred measure for SWPW and it is also the best measure for the other industries examined in this chapter. Neither it nor the equipment series perform satisfactorily in the sawmill supply equation. The resulting regressions often do not converge and statistical tests on their cost of capital and the wage coefficients are most unsatisfactory. The new series for structures, however, performs very well. Perhaps the explanation is that longer term investments in fixed plant are particularly important for the sawmill industry during the 1950-1980 period. This would be consistent with general observations of rapidly increasing plant scale together with decreasing industry-wide employment of labor. It is also consistent with Buongiorno and Lu's (1989) observation that sawmill investments are small and better predicted by lumber price than by the general cost of capital.

Final estimates of equation (5.1) omit the proxy for private R&D expenditures. The sawmill industry is most competitive and many mills are small operations. Therefore, it is unlikely that many firms in the industry conduct their own independent R&D. Rather, the government and the intermediate industries that produce sawmill equipment develop any new equipment. The sawmill industry simply buys new technology when it buys new capital equipment. Both conversations with researchers at the Forest Products Laboratory (FPL) and subsequent statistical tests support this contention—and this formulation of the sawmill supply equation.[3]

The FPL supports all public sawmill research. Table 5.1 records the expenditure history of this research for 1950-1980. The correct research-implementation lag is the important remaining R&D characteristic. The lag between public sawmill research and its initial impact on output is five years.[4] This sawmill lag is longer than that for the other forest products industries that we examine in this book. (The comparable SWPW lag was only two years.) The highly competitive nature of the sawmill industry, therefore its reliance on its equipment suppliers for new technologies, anticipates this result. It is reasonable that technological development and marketing by the intermediate, equipment producing, industry may take awhile. Nevertheless, some public R&D does not require new equipment for implementation. Improved lumber scaling techniques and early studies of log cutting geometry (later computerized in the BOF project) are examples. Even in these cases, however, the large number of firms in the sawmill industry (approximately 6800 in 1980 compared with only 80 in the SWPW industry) and the small size of a great many of them insures that complete information transfer is slow.

The demand function calls on knowledge that the demand for lumber derives from the demands for new housing and for home improvement.

$$q_t = \beta_0 + \beta_1 p_t + \beta_2 hs_t + \beta_3 q_{h,t} + \beta_4 p_{s,t} \tag{5.2}$$
$$(-) \quad (+) \quad (?) \quad (+)$$

All variables are in logarithmic form. The definitions of previously unidentified variables and their data sources are

Table 5.1. FPL effort in sawmill and planing mill research 1950-1980 (in 1967$)

year	scientist months	cost per scientist month[*]	total cost
1950	57	$ 5204.54	$ 296,658.78
1951	56	4855.78	271,923.68
1952	60	5115.39	306,923.40
1953	56	5251.19	294,066.64
1954	54	5313.59	286,933.86
1955	54	5420.57	292,710.78
1956	36	5393.55	194,167.80
1957	38	5360.01	203,680.38
1958	39	5373.36	209,561.04
1959	35	5491.13	192,189.55
1960	30	5612.64	168,379.20
1691	22	5736.43	126,201.46
1962	23	5854.13	134,644.99
1963	36	6017.04	216,613.44
1964	35	6156.62	215,481.70
1965	45	6222.45	280,010.25
1966	40	6246.15	249,846.00
1967	46	6491.89	298,626.94
1968	33	6609.92	218,127.36
1969	58	6716.05	389,530.90
1970	48	6861.94	329,373.12
1971	63	6974.01	439,362.63
1972	58	7029.79	407,727.82
1973	77	6735.11	518,603.47
1974	63	6231.16	392,563.08
1975	129	6396.18	825,107.22
1976	116	6266.84	726,953.44
1977	107	6229.98	666,607.86
1978	86	6269.30	539,159.80
1979	108	5964.67	644,184.36
1980	97	5457.57	529,384.29

[*] Includes overhead
Source: FPL attainment reports

hs = new housing starts (Ulrich)

q_h = a quantity index of hardware store sales (provided by R. Boyd, Ohio University)

p_s = a price index for related goods (provided by R. Boyd, Ohio University)

q_h is a proxy for the level of home improvement. Boyd derived this index by dividing total hardware store sales (US Department of Commerce 1983) by the consumers' price index to obtain a measure of quantity, and then divided this measure in each year by the 1967 quantity. Boyd's price index for related goods is a Divisia index of real prices for concrete, structural steel, plywood and veneer, and gypsum. Each of these are materials used, together with lumber, in housing construction. The weights for each material originate with input-output tables for the US. Boyd used linear interpolations to construct weights for unreported years.

The parenthetical expressions below the demand equation show the anticipated signs for the unknown coefficients. The anticipated own price elasticity β_1 is negative. The anticipated sign on β_3 depends on whether related goods are net substitutes or net complements with lumber in consumption.

Table 5.2 records the non-linear two-stage least-squares (NL2SLS) econometric estimates. The supply coefficients all have the expected signs

Table 5.2. NL2SLS estimates of demand and supply coefficients for sawmills and planing mills

	supply		demand
labor (α_1)	0.1698** (0.0747)	(β_0)	11.5677** (4.9917)
capital (α_2)	0.0823* (0.0302)	own price (β_1)	-0.6003*** (-0.4552)
G_{t-5} (μ)	0.0280** (0.0149)	housing starts price (β_2)	0.2938** (0.1582)
lag (λ)	0.9698* (0.0235)	hardware store sales (β_3)	0.5588** (0.2892)
		related goods price (β_4)	-0.0718**** (-0.0752)
R^2	.67		.75
Durbin's h	0.298		---
Durbin-Watson	---		1.269
degrees of freedom	22		21

Numbers in parentheses are standard error.
  * Significant at the 1 percent level in a one-tailed test
 ** Significant at the 5 percent level in a one-tailed test
*** Significant at the 10 percent level in a one-tailed test
**** Significant at the 20 percent level in a one-tailed test

and all are significant at the 5 percent level or better.[5] The output elasticities for labor and capital are 0.17 and 0.08, respectively. The output elasticity for public R&D is $\mu/(1 - \lambda) = 0.93$. It is significant at the 5 percent level.[6]

The output elasticities should sum to one—or less, if there are significant scale economies. Therefore, the hypothesis that $1 - \alpha_1 - \alpha_2 - \mu/(1 - \lambda) = 0$ is a test for our exclusion of a private R&D proxy in the supply equation. Our estimate of one minus the output elasticities is -0.18 with a standard deviation of 0.38. Thus, we cannot reject the hypothesis that the output elasticities sum to one. Neither can we reject the exclusion of private R&D from the supply equation.

Comparing our supply price elasticity with prior econometric estimates confirms our confidence in the table 5.2 results. Lewandrowski (1989) made the most thorough assessment of lumber supply and demand known to us. Adams and Haynes (1980) built the best-known macroeconomic model of the forestry sector of the US economy. They find regional supply price elasticities in the ranges of 0.18-0.38 and 0.21-0.79, respectively. Our supply price elasticity is $(\alpha_1 + \alpha_2)/(1 - \alpha_1 - \alpha_2) = 0.34$ and is significant at the 1 percent level. This estimate falls within the narrow combined range of Lewandrowski's and Adams-Haynes estimates.

The demand coefficients have the anticipated signs and they are generally significant at the 1 percent level. The negative coefficient on related goods β_4 suggests that related goods are net complements. This coefficient is significant only at the 20 percent level. Nevertheless, theory justifies its presence in our equation and excluding it might bias the other coefficients.

The demand price elasticity of -0.60 is significant at the 10 percent level. Lewandrowski observes regional elasticities between -0.27 and -0.44. Our aggregate elasticity is greater than these regional estimates, but it is in the neighborhood of Robinson's (1974) estimate of -0.87 for the entire US.

Results: The Efficiency of Public Sawmill Research

Public research induces an associated private effort where specialized applications are necessary for the adoption of new public technologies. We have no general estimate of the magnitude of the implementation effort for the sawmill industry. Rather, as in the SWPW industry, we obtain a measure of the induced private effort from knowledge of a representative case, and then test our eventual research benefit estimates for sensitivity to variations in this measure.

There are various examples of new techniques and new equipment developed by the Forest Products Laboratory. Our example is the BOF equipment. The original idea for developing a computer program to determine the optimal cutting angles for any particular log was Hiram Hallock's. David Lewis worked with Hallock. Lewis and Hallock's original discussions took place around 1969 and Lewis eventually did most of the research in the early 1970s. Lewis reviewed these events with us, recalling the various steps in the development of the BOF equipment and determining that 39 scientist-months were spent on the project at the FPL before private firms took over.[7] At the 1971 cost of R&D, 39 scientist-months equal $271,986 (1967$).

The BOF technology is freely available from the FPL for all firms that choose to enter the BOF equipment industry. Therefore, entry must occur into this industry until economic profits fall to zero and there are no further development costs after the FPL activity. Only sawmill implementation costs remain.[8]

BOF equipment can be relatively expensive for the many smaller sawmills. Therefore, only 50-60 percent of all sawmills have adopted any part of the technology. Adoption ranges from the purchase of tables identifying the BOF to full operation of computer, setting or scanning equipment. The cost range is $50-500,000 (1989$), but the cost distribution is bimodal with one mode at $50-75,000 and the other at approximately $250,000.[9] This converts, in 1967 dollars, to a range of induced private costs per public research dollar between $0.06 and $0.58 per sawmill—or $0.03-0.29 per sawmill if one-half of all mills adopt the technology. The average of the two investment modes converts to an induced private cost multiplier (hereafter, the "multiplier") of $0.09 per sawmill if one-half of all mills adopt the technology.

Table 5.3 shows the estimated total sawmill R&D costs, public and private, using three alternative multipliers. The multipliers are small (approximately one-half those for SWPW), indicating a small ratio of implementation costs per mill to FPL research costs. Nevertheless, total R&D costs are large (2-3 orders of magnitude larger than total annual SWPW costs), in part because of the large FPL expenditure for sawmill research (approximately an order of magnitude larger than annual FPL expenditures on SWPW research), but largely because there are so many sawmills.

Table 5.4 shows the net present value (NPV) estimates; benefit-cost (BC) ratios; and internal rates of return (IRR) for consumers, producers and for society as a whole for the 1950-1980 program of public sawmill research. The NPV and BC estimates reflect sensitivity to the range of social discount rates between 4 and 10 percent. Table 5.5 shows the long-run value of the marginal product (VMP) for each individual year's investment, the average long-run VMP, and the marginal internal rate of return (MIRR).

Table 5.3. Social cost of public research in sawmills, 1950-1980 (direct cost plus cost of adoption, in millions of 1967$)

year	number of plants	multiplier		
		0.03	0.09	0.29
1950	18,867	168.6	505.1	1,626.9
1951	18,968	156.1	467.7	1,506.5
1952	19,070	175.9	527.0	1,697.4
1953	19,171	168.5	505.0	1,626.7
1954	19,273	165.7	496.5	1,599.3
1955	18,364	160.7	481.6	1,551.1
1956	17,455	102.0	305.6	984.2
1957	16,546	101.8	305.0	982.4
1958	15,637	98.0	293.5	945.2
1959	14,947	85.6	256.5	826.1
1960	14,258	72.0	215.5	694.1
1961	13,568	52.1	156.1	502.6
1962	12,879	52.6	157.6	507.6
1963	12,189	78.8	235.9	759.5
1964	11,710	75.4	225.7	726.7
1965	11,230	94.8	284.0	914.4
1966	10,751	81.7	244.5	787.4
1967	10,271	93.9	179.3	899.3
1968	9,831	64.9	194.4	625.9
1969	9,391	109.3	327.1	1,053.2
1970	8,951	88.3	264.2	850.5
1971	8,511	113.2	338.6	1,090.0
1972	8,071	98.3	294.2	947.0
1973	7,966	125.2	374.6	1,205.7
1974	7,860	93.5	279.7	900.3
1975	7,755	193.2	578.0	1,860.6
1976	7,649	166.3	497.4	1,601.2
1977	7,544	151.2	542.2	1,455.5
1978	7,298	119.2	356.6	1,148.0
1979	7,053	137.5	411.1	1,323.4
1980	6,087	109.0	326.0	1,049.5

The *NPV* and *BC* estimates reflect sensitivity to the range of induced private cost multipliers and to the range of social discount rates between 4 and 10 percent. All calculations derive from knowledge of the price elasticities, R&D elasticities, and the research-implementation lag in the supply and demand equations and from the formulae for the *NPV*, *IRR*, *VMP*, and *MIRR* derived in chapter 3.

The positive *NPV*s reported in table 5.4 imply that public sawmill research for the period 1950-1980 produced net gains for both consumers and society as a whole. Producers would be net losers if they had to fund all sawmill research themselves. The price decreases associated with expanding markets, together with the research costs themselves, more than offset any increase in total revenues due to technology-induced falling production costs. In fact, producers seldom do invest in sawmill research. This observation, together with our observation of new social gains, supports the historic public role in sawmill research.

Table 5.4. Returns to public investment in sawmill research, 1950-1980 (millions of 1967$)

multiplier		social discount rate		
		.04	.07	.10
.03	NPV^{cs}	83,694.3	42,000.9	23,132.0
	NPV^{ps}	-27,703.8	-14,529.9	-8,463.7
	NPV^{neb}	58,057.6	28,991.2	15,860.8
	BC^{cs}	39.89/1	26.76/1	18.55/1
	BC^{neb}	29.97/1	18.76/1	13.00/1
	IRR^{cs} - 57%			
	IRR^{neb} - 49%			
.09	NPV^{cs}	79,570.9	38,968.0	20,752.6
	NPV^{ps}	-31,827.2	-17,562.8	-10,843.2
	NPV^{neb}	53,934.2	25,958.2	13,481.3
	BC^{cs}	13.22/1	8.93/1	6.19/1
	BC^{neb}	9.34/1	6.25/1	4.35/1
	IRR^{cs} - 34%			
	IRR^{neb} - 28%			
.29	NPV^{cs}	65,826.4	28,858.1	12,821.0
	NPV^{ps}	-45,571.7	-27,672.7	-18,774.9
	NPV^{neb}	40,189.7	15,848.4	5,549.6
	BC^{cs}	4.14/1	2.77/1	1.92/1
	BC^{neb}	2.90/1	1.94/1	1.35/1
	IRR^{cs} - 17%			
	IRR^{neb} - 13%			

The *IRR* ranges from 13 to 57 percent depending on the preferred implementation cost multiplier. This observation is consistent with the Haygreen *et al.* (1986) forecast of 36 percent for benefits to the year 2000 from various sawmill research projects conducted between 1972 and 1982. Haygreen's observations and ours fall in the low-to-middle range of experience for agriculture research. The *MIRR* estimates of 5 to 31 percent suggest that public research was equal to or greater than the socially optimal level, the level that would produce social returns approximating the marginal social opportunity cost of capital.

These results are satisfying for public research managers, although they are less impressive than the comparable results for SWPW. There may be three reasons: (1) The five-year research-implementation lag implies larger discounting of sawmill research benefits. (2) The larger number of sawmills means larger private implementation costs. (3) There was a larger expenditure of public funds for sawmill research. Softwood plywood research was a smaller scale activity and it existed for a briefer time period. It may have always been under greater management scrutiny, while sawmill research was more established. The sawmill research budget may have been more secure and the large number of sawmills, at least one in almost every Congressional District, may have insured its political support.

Table 5.5. Long-run VMPs and MIRRs of public sawmill research, 1950-1980 (1967$)

multiplier	0.03 social discount rate			0.09 social discount rate			0.29 social discount rate		
year	.04	.07	.10	.04	.07	.10	.04	.07	.10
1950	8.04	5.03	3.46	2.68	1.68	1.16	0.83	0.52	0.36
1951	7.90	4.94	3.41	2.64	1.65	1.14	0.82	0.51	0.35
1952	7.26	4.54	3.13	2.42	1.51	1.04	0.75	0.47	0.32
1953	7.67	4.79	3.30	2.56	1.60	1.10	0.79	0.50	0.34
1954	7.55	4.72	3.25	2.52	1.58	1.09	0.78	0.49	0.34
1955	8.47	5.30	3.65	2.83	1.77	1.22	0.88	0.55	0.38
1956	13.03	8.15	5.61	4.35	2.72	1.87	1.35	0.84	0.58
1957	11.12	6.95	4.79	3.71	2.32	1.60	1.15	0.72	0.50
1958	10.73	6.71	4.62	3.58	2.24	1.54	1.11	0.70	0.48
1959	14.16	8.85	6.10	4.73	2.96	2.04	1.47	0.92	0.63
1960	15.24	9.53	6.57	5.09	3.18	2.19	1.58	0.99	0.68
1961	19.60	12.26	8.44	6.54	4.09	2.82	2.03	1.27	0.88
1962	20.18	12.62	8.69	6.74	4.21	2.90	2.09	1.31	0.90
1963	14.46	9.04	6.23	4.83	3.02	2.08	1.50	0.94	0.65
1964	15.95	9.97	6.87	5.33	3.33	2.29	1.65	1.03	0.71
1965	12.47	7.80	5.37	4.17	2.60	1.79	1.29	0.81	0.56
1966	14.18	8.87	6.11	4.74	2.96	2.04	1.47	0.92	0.63
1967	12.81	8.01	5.52	4.28	2.68	1.84	1.33	0.83	0.57
1968	20.83	13.02	8.97	6.96	4.35	3.00	2.16	1.35	0.93
1969	12.92	8.08	5.57	4.32	2.70	1.86	1.34	0.84	0.58
1970	13.88	8.68	5.98	4.64	2.90	2.00	1.44	0.90	0.62
1971	12.59	7.87	5.43	4.21	2.63	1.81	1.31	0.82	0.56
1972	18.88	11.80	8.13	6.31	3.95	2.72	1.96	1.23	0.84
1973	15.90	9.94	6.85	5.31	3.32	2.29	1.65	1.03	0.71
1974	16.78	10.49	7.23	5.61	3.51	2.42	1.74	1.09	0.75
1975	6.69	4.19	2.88	2.24	1.40	0.96	0.70	0.43	0.30
1976	9.80	6.13	4.22	3.28	2.05	1.41	1.02	0.64	0.44
1977	12.62	7.89	5.44	4.22	2.64	1.82	1.31	0.82	0.56
1978	17.09	10.69	7.36	5.72	3.57	2.46	1.78	1.11	0.76
1979	14.73	9.21	6.34	4.92	3.08	2.12	1.53	0.96	0.66
1980	14.46	9.17	6.32	4.91	3.07	2.11	1.52	0.95	0.66
average VMP	12.55	7.85	5.41	4.19	2.62	1.81	1.30	0.81	0.56
MIRR	31%			16%			5%		

Nevertheless, it should be clear that sawmill research (a) would not have occurred without active public involvement, (b) the public involvement produced positive social returns and (c) more research would have produced greater social returns yet. Sawmill research was a socially productive investment of public funds.

WOODPULP RESEARCH

Wood fiber became an important input in paper manufacturing in the later nineteenth century. From the start, woodpulp mills in the US followed in the geographic path of lumber mills. That is, pulp production initially concentrated in New England, but eventually, by the 1960s, it shifted southward. As lumber production declined in relative importance in the South, pulp production increased. By 1970, the South produced 63 percent of the national output of 48 million tons—worth $1,142.7 million (1967 dollars) annually (McKeever 1987).

By the mid-1970s there were 26 firms with 324 mills in the US producing woodpulp of all varieties (McKeever 1987, Bureau of the Census 1985b). The four largest firms controlled 48 percent of the market but only 13-15 percent of total woodpulp production traded on the open market. Many firms are vertically integrated to include both pulp and papermills. These firms treat pulp as an intermediate output in paper production and transfer their pulp production directly to their own papermills without reference to the market. This information, altogether, suggests a potentially non-competitive market—for which we would have difficulty estimating an industry supply function.

Nevertheless, if profit-maximizing vertically-integrated paper companies price internally-produced woodpulp at its marginal cost, then independent pulp producers cannot sell their own output to the paper companies at prices much different from their own marginal costs. Higher prices charged by the independent producers would induce the integrated producers to expand internal production. Therefore, marginal cost pricing must rule the industry and an industry supply function can be defined from the dual.

Public research for the woodpulp industry through the 1950s focused on the pulping and bleaching process. By the late 1950s, however, it began changing to reflect the environmental concerns of the time. The pulping process uses large volumes of water and leaves a low oxygen effluent. The FPL began research featuring oxygen pulping and polysulfide pulping methods as means of decreasing pollution. Neither method was widely adopted and the FPL never got involved in the alternative approach of pollution control devices. Research in the later 1970s changed again to reflect the energy crisis. An FPL task force, in response to a higher government mandate, examined best available technologies for producing alcohol from wood. Researchers thought this an unlikely prospect for lowering energy costs and their expectations proved correct. Wood alcohol research was an unproductive venture.[10]

Pollution abatement research may not be reflected in increased production. The politically motivated, and unproductive, energy research certainly created no technological improvements. These activities imposed research costs but yielded no social benefits. Therefore, the inclusion of their costs in our analysis will decrease the final net impacts of aggregate public woodpulp research. The costs will make our estimates of net gains from strictly FPL initiated research conservative.

Supply and Demand

The supply function, comparable to equation (3.7) with its (small and constant) bracketed term removed, is

$$q_t = (1 - \lambda)\ln A + \gamma(\alpha_1 + \alpha_2)(p_t - \lambda p_{t-1}) - \gamma\alpha_1(w_t - \lambda w_{w-1})$$
$$-\gamma\alpha_2(r_t - \lambda r_{t-1}) + \gamma\eta h_{t-2} + \gamma\mu g_{t-3} + \lambda q_{t-1} \qquad (5.3)$$

where γ and A are constant functions of α_1 and α_2. All exogenous variables are in logarithmic form. Their definitions and their data sources are

q = quantity of woodpulp in thousands of short tons (Ulrich 1983)
p = price per thousand short tons (Ulrich 1983 and *The Pulp and Paper Trade Journal* 1978), deflated by the producers' price index
w = average hourly wage for production workers in pulpmills (US Department of Labor 1979, 1983), deflated by the producers' price index
r = real user cost of capital for paper and allied products (Wharton Econometrics, personal correspondence)
h = (a proxy) private expenditures on R&D in paper and allied products (National Science Foundation 1981)[11]
g = government scientist months for woodpulp research (various FPL attainment reports)

The price term and the private and public R&D terms require more careful explanation. Prices vary for different grades and they are not easily available to industry outsiders. Our solution relies on Ulrich (1983) and the industry trade journal. Ulrich provides a producer price index for pulp. *The Pulp and Paper Trade Journal* provides the price of market kraft pulp in 1976. The kraft price can be converted to 1967 dollars using the all commodities producer price index. This provides a 1967 base point for adjusting the rest of the series.

The woodpulp supply formulation uses a direct proxy for private R&D—rather than the unknown constant fraction of total revenue used as a proxy for private R&D in the SWPW analysis in chapter 4. Finally, the lag between public research and its initial impact on production is three years in the woodpulp case. (The process for determining this lag is the same as for the SWPW and the sawmill research-implementation lags.)

The demand function is more problematic. Paper, the consuming industry for pulp, is a high fixed cost industry that is slow to respond to cyclical variations in final demand. It has run at 89-96 percent of capacity for all but two years since 1950 (McKeever 1987). This suggests that papermill capacity is the primary determinant of pulp demand and that the demand price elasticity is close to zero. Indeed, Gilless and Buongiorno's (1987) definitive model treats pulp and paper as a fixed coefficient industry with aggregate industry capacity as the only determinant of the demand for pulp and Guthrie (1972) finds a demand price elasticity of −0.005.

Our problem lies in the data series for papermill capacity. The technical efficiency of pulp conversion to paper has improved since 1950, but we have no measure of this improvement. Without adjustment for improved efficiency, the capacity series is unreliable.

A less-than-satisfactory alternative is to estimate demand in a manner consistent with the chapter 4 estimate for softwood plywood. This argues for a demand function with terms for own price, the costs of labor and capital and substitute inputs in the consuming industry, and final good price.

$$q_t = \beta_0 + \beta_1 p_t + \beta_2 w_{p,t} + \beta_3 r_t + \beta_4 c_t + \beta_5 p_{p,t} \atop \ \ \ \ \ \ \ \ \ (-) \ \ \ \ \ (+) \ \ \ \ \ \ (+) \ \ \ \ \ (+) \ \ \ \ \ (+) \tag{5.4}$$

All variables are in log form. The definitions of the new exogenous variables and their data sources are

p_p = index of paper price (Ulrich 1983)
w_p = average hourly wages for paper production workers (US Department of Labor 1979, 1983)
c = index of chemical price (US Department of Commerce 1983).

All other variables remain as defined in the supply function. All price and cost data for both the supply and demand equations are converted to 1967 units using the all commodities producer price index. User cost of capital data are available only for the combined paper and allied products industry. Therefore, our cost of capital data are identical for the producing industry in the supply equation and for the consuming industry in the demand equation.

The parenthetical expressions below the demand equation reflect the anticipated signs for the unknown coefficients. The user cost of capital and the final good price are determinants of capacity. Increasing final good prices and falling capital costs induce capacity expansion. Therefore, these variables may act as instruments for a capacity term in the demand equation. The cost of capital data are composites of short- and long-run rates. Therefore, they may also reflect any (anticipated minimal) effect of variable capital: chemicals in this case. For this reason, and because only capacity and not variable capital may be determinants of demand, we evaluate equation (5.4) both with and without the chemical price index,

$\beta_4 = 0$ and $\beta_4 \neq 0$, respectively.

Table 5.6 shows the NL2SLS estimates for both formulations. The supply coefficients all have the expected signs and all except the government R&D coefficients are significant at the 5 percent level or better. The government R&D coefficient is significant at the 10 percent level in the system including chemicals ($\beta_4 \neq 0$) and at the 20 percent level in the system excluding chemicals ($\beta_4 = 0$). These alternative government R&D coefficients are within one standard deviation of each other. Durbin's *h* statistic indicates possible serial correlation in both equations. The coefficient estimates are (unbiased and) more significant than the *t* statistics indicate since autocorrelation causes standard errors to be biased downward in absolute value. In sum, the supply coefficients are reliable for our purposes.

Choice between the two equations has almost no impact on the important elasticities for our subsequent research evaluation. The system excluding chemicals yields slightly more conservative research benefits. The supply price elasticity of 1.09 is significant at the 1 percent level in this system. The long-run public research output elasticity of 0.09 is significant at the 10 percent level.

The demand coefficients are not so reliable. The coefficients on wages

Table 5.6. NL2SLS estimates of demand and supply coefficients for woodpulp

	supply			demand	
	$\beta_4=0$	$\beta_4 \neq 0$		$\beta_4=0$	$\beta_4 \neq 0$
labor (α_1)	0.3057* (.0650)	0.2875* (.0664)	β_0	8.9160* (2.3450)	0.5182 (11.4200)
capital (α_2)	0.2164* (.0484)	0.2233* (.0492)	own price (β_1)	1.0140* (0.3540)	3.0910**** (2.4330)
H_{t-2} (n)	0.2450* (.0751)	0.2552* (.0766)	labor (β_2)	1.8460* (.2099)	1.9170* (.4576)
G_{t-3} (μ)	0.0686*** (.0548)	0.0732*** (.0556)	capital (β_3)	-.3840* (-.1516)	0.3659 (0.5250)
lag λ	0.2388** (.1173)	0.2356** (.1213)	chemical price (β_4)	----	-2.0280**** (-1.8240)
			paper price (β_5)	-1.5500** (-.8210)	3.1839**** (-2.4160)
R^2	.87	.87		.57	.83
Durbin's h	2.83	2.89		--	--
Durbin-Watson	--	--		1.10	.66
degrees of freedom	23	23		23	22

Numbers in parentheses are standard errors.
 * Significant at the 1 percent level
 ** Significant at the 5 percent level
 *** Significant at the 10 percent level
 **** Significant at the 20 percent level

and the cost of capital conform with expectations. The latter is more important because it is more closely associated with our expectations about capacity as a determinant of demand. It is significant at the 1 percent level in the more favored case where the cost of capital coefficient also attracts any impact of variable capital ($\beta_4 = 0$). Nevertheless, the own price elasticity β_1 is positive in both cases—contrary to the law of demand. And the coefficient on paper price should be positive (as paper price increases, the demand for pulp should increase). Furthermore, the Durbin-Watson statistic for neither specification provides confidence in these demand equations.

This unsatisfactory demand formulation, together with our prior preference for a form emphasizing papermill capacity, force us to turn to external evidence for the measure of pulp demand price elasticity. Gilless and Buongiorno (1987) encourage the assumption of perfectly inelastic demand ($\beta_1 = 0$). This is our assumption in the subsequent calculations of research benefits. Perfectly inelastic demand implies more conservative estimates of final research benefits than Guthrie's (1972) ever-so-slightly elastic demand estimate ($\beta_1 = 0.005$).[12]

Results: The Efficiency of Public Woodpulp Research

Woodpulp research in the period from 1950 to 1980 involved improvements in existing technologies, but did not generally require new equipment. Some research, like that on polysulfide pulping, would have required new equipment if it had been implemented. Furthermore, current research on biopulping will require new equipment if the research is successful and the resulting new technology is adopted. Nevertheless, the public R&D improvements of 1950-1980 did not generally require new equipment. Therefore, the costs for each pulpmill of adopting acceptable FPL research were negligible. The essential social cost of R&D is the direct cost of FPL research. Table 5.7 shows costs per scientist month and total annual costs for the full period.

Table 5.8 shows the estimated net present value (*NPV*), benefit/cost (*BC*) ratios, and internal rates of return (*IRR*) for consumers, producers, and for society as a whole for this thirty-year program of public woodpulp research. The *NPV* and *BC* estimates reflect sensitivity to the range of social discount rates between 4 and 10 percent. Table 5.9 shows the long-run value of the marginal product (*VMP*) for each individual year, the average long-run *VMP* and the marginal internal rate of return (*MIRR*). All calculations derive from knowledge of the price elasticities, R&D elasticities, and the research-productivity lag in the supply and demand equations and from the formulae for the *NPV, IRR, VMP,* and *MIRR* derived in chapter 3.

Table 5.7. FPL effort in woodpulp research, 1950-1980 (in 1967$)

year	scientist months	cost per scientist month*	total cost
1950	97.25	$ 5204.54	$ 506,141.43
1951	151.25	4855.78	734,436.85
1952	168.75	5115.38	863,221.22
1953	141.25	5251.19	741,731.18
1954	132.00	5313.59	701,393.44
1955	119.75	5420.57	649,113.26
1956	111.50	5393.55	601,381.29
1957	107.00	5360.01	573,521.43
1958	99.50	5373.36	534,649.15
1959	112.60	5491.13	618,301.61
1960	100.00	5612.67	561,267.00
1961	97.00	5736.43	556,433.79
1962	95.00	5854.13	556,142.51
1963	116.10	6017.04	698,578.34
1964	168.00	6156.62	1,034,311.46
1965	179.50	6222.45	1,116,929.63
1966	218.50	6246.15	1,364,783.23
1967	208.00	6491.89	1,350,313.47
1968	172.00	6609.92	1,136,906.67
1969	128.80	6716.05	865,027.13
1970	112.30	6861.94	770,596.14
1971	124.80	6974.01	870,356.24
1972	139.60	7029.79	911,061.11
1973	115.20	6735.11	775,884.19
1974	122.00	6231.16	760,201.21
1975	195.75	6396.17	1,252,051.26
1976	147.50	6226.84	918,458.53
1977	142.20	6229.98	885,903.75
1978	192.30	6269.30	1,205,586.23
1979	174.90	5964.67	1,043,220.49
1980	174.00	5457.57	949,617.76

* Includes overhead
Source: FPL attainment reports

Table 5.8. Returns to public investment in woodpulp research, 1950-1980 (in millions of 1967$)

	social discount rate		
	0.04	0.07	0.10
NPV[cs]	33.917	18.049	10.208
NPV[ps]	-36.849	-22.749	-15.203
NPV[neb]	11.034	4.856	1.991
BC[cs]	3.30/1	2.70/1	2.24/1
BC[neb]	1.72/1	1.41/1	1.17/1

IRR[cs] — 33%
IRR[neb] — 15%

Table 5.9. Long-run VMPs and MIRRs of public woodpulp research, 1950-1980

year	social discount rate		
	0.04	0.07	0.10
1950	$ 0.32	$ 0.30	$ 0.27
1951	.27	.24	.22
1952	.23	.21	.19
1953	.28	.26	.23
1954	.31	.28	.26
1955	.39	.35	.32
1956	.45	.41	.38
1957	.46	.42	.38
1958	.49	.45	.41
1959	.48	.43	.40
1960	.54	.49	.45
1961	.55	.50	.45
1962	.57	.52	.48
1963	.48	.43	.40
1964	.36	.33	.30
1965	.35	.32	.29
1966	.30	.27	.25
1967	.30	.28	.25
1968	.39	.36	.33
1969	.52	.47	.43
1970	.63	.57	.52
1971	.55	.50	.46
1972	.54	.49	.45
1973	.66	.60	.55
1974	.97	.88	.80
1975	.62	.57	.52
1976	.91	.83	.75
1977	.90	.82	.75
1978	.59	.54	.49
1979	.73	.67	.61
1980	.81	.74	.68
average VMP	.48	.44	.40

MIRR < 0   (average VMP at zero discount rate = 0.55)

Both consumers and society as a whole obtain net gains from public woodpulp research. The perfectly inelastic demand for woodpulp insures that pulp producers must be net losers. This result suggests that there is little private incentive to invest in R&D and considerable incentive to vertically integrate pulpmills with papermills in order to capture the surpluses from successful public research. Indeed, we observed at the beginning of this pulpmill discussion that the industry is largely vertically integrated. An open market accounts for only 13 percent of all woodpulp transactions.

The positive net economic benefits reported in table 5.8 imply that public woodpulp research between 1950 and 1980 produced net gains for society. The *IRR* for woodpulp research is in the range of 15-33 percent. This falls in the lower end of the range of experience for agricultural research. Furthermore, the negative *MIRR* suggests overinvestment. That is, reduced investment in woodpulp research in this period would have caused the *NPV* and the average *IRR* to increase. In sum, the net gain was positive but society would have been even better off with a lower level of total investment. This *MIRR* is consistent with observations of the staff at the Forest Products Laboratory that many of their innovations were never widely adopted and that some research investment was perhaps a misguided political imposition.

These results raise questions about the continued high level of public woodpulp research over the years regardless of the presence of low research adoption rates and other indications of research inefficiency. One response is that there was a prior (pre-1950) experience of research success and adoption. This experience holds hope for the return to former levels of research success. Indeed, marginal returns have been increasing since 1970. Another response is that the FPL is a public agency. It may experience difficulty modifying its research staff size in the short run. Furthermore, its research budget may have been encouraged by a highly visible industry interest group and its research activities have been directed, at times, by political interest rather than researcher anticipation of technical success or likelihood of adoption.

There is also another, altogether different, and plausible explanation for the low *MIRR* for public woodpulp research. Much of the research in the late 1960s was designed to continue current production levels while altering the flow of polluting effluent. To the extent that this pollution control research was adopted, yet was not cost-reducing but instead altered the production process at unchanging output levels, then its impact is not reflected in our benefit estimation. This implies that the overall *NPV, IRR,* and *MIRR* all underestimate of the social gain. Apparently, the remaining research, other than for pollution control, had a sufficiently large cost-reducing impact to offset some uncompensated pollution control research costs, thereby creating aggregate measures for all woodpulp research reflecting a positive net economic benefit and an average internal rate of return exceeding 15 percent.

WOOD PRESERVATIVES RESEARCH

The four-firm concentration ratio for the wood preservatives industry was in the 30 to 40 percent range throughout our period of analysis. This suggests a competitive industry and permits us to define its supply/production dual.

Wood preservatives extend the life of treated wood products. Therefore, one effect of wood preservatives research is on product quality. Improved quality benefits consumers but it is not the cost-reducing (process innovation) research reflected in our measure of technical change. Furthermore, wood preservatives are generally petroleum products. The residuals created while treating wood with petroleum products are environmentally objectionable. Therefore, much recent research in the wood preservatives industry has the objective of reducing levels of environmentally damaging residuals while continuing to produce known products. These changes in the residual product are also hidden from our output measure.

Our eventual estimates of research productivity in the wood preservatives industry must be underestimates by the magnitude of these (a) product quality and (b) environmental research impacts. Since we anticipate that product quality and environmental research components are large in this industry relative to other forest products industries, then we anticipate that the *observed* level of research productivity in the wood preservatives industry may be a lower bound for *total* levels of research productivity in all forest products industries.

Historical Background

Wood preservatives research has been an important activity of the Forest Products Laboratory from its beginning in 1910. One of the FPL's four original divisions dealt with wood preservatives. Initially, railroad crossties were the predominant market for wood treatment (90 percent of total volume). Plentiful inventories of naturally durable chestnut and cedar were available for poles, the common alternative use of treated wood.

The blight destroyed the chestnut inventory and the accessible cedar inventory was gradually consumed. Meanwhile, rural electrification projects increased the demand for treated telephone poles and crossties and the shift in southern agriculture from cotton to cattle increased the demand for treated fence posts. The demand for commercially produced and treated poles increased further yet as farms became less self-sufficient. Moreover, the original variations in quality among commercially produced poles and posts, together with increasing farm demand, created a new consumer demand for industry-wide specifications. The FPL has played a large role since the 1950s in setting specifications and developing methods for testing wood preservative treatments.

The major innovation of the 1960s was the dual treatment (particularly for marine pilings) of chromate copper arsenate and creosote. The chromate copper arsenate treatment was new in the 1960s. It remains the dominant inorganic preservative today. Creosote has always been an important preservative. The dual treatment process reduced susceptibility to attack by marine organisms but it created brittle pilings. Therefore, it is not a fully successful innovation.

The energy crisis and rising environmental concerns focused research in the 1970s on decreasing the use of petroleum products and decreasing environmental residuals, respectively. These factors, together with the decline of the railroad industry and an increasing demand for treated wood in residential construction, encouraged a three-fold expansion in the use of inorganic arsenicals between 1970 and 1980. Railroad ties declined from a 48 percent wood preservatives market share in 1950 to a 31 percent share in 1980 and poles declined from a 29 percent to a 20 percent share while treated lumber and timber increased from a 12 percent to a 35 percent share over the same period. These shifts altered the FPLs foci in the direction of inorganic preservatives and new specifications for treated lumber and timber. Table 5.10 shows the pattern of FPL research expenditures over this period.[13]

Supply and Demand

The supply function, comparable to equation (3.7), is

$$
\begin{aligned}
q_t = (1 - \lambda)\ln A &+ \gamma(\alpha_1 + \alpha_2)(p_t - \lambda p_{t-1}) - \gamma\alpha_1(w_t - \lambda w_{t-1}) \\
&- \gamma\alpha_2(r_t - \lambda r_{t-1}) + \gamma[\theta t - \lambda\theta(t-1)] + \gamma\eta h_{t-1} \qquad (5.5) \\
&+ \gamma\mu g_{t-1} + \lambda q_{t-1}
\end{aligned}
$$

where the definitions of A and γ are unchanged. The small and constant bracketed term in equation (3.7) is removed, as it was in all previous empirical cases, but a term for disembodied technical change has been added.[14] The inclusion of disembodied technical change is motivated by entirely statistical considerations. The coefficient on disembodied technical change θ was insignificant for previous industries. It is significant and negative in the wood preservative case and the wood preservative equations do not converge when it is removed. Because θ is negative, it probably accounts for some unidentified trend different from technical change.

All exogenous variables are in logarithmic form. Their definitions and their data sources are

q = a volume measure of preserved wood products (US Department of Commerce *Census of Manufactures*, various years)

p = own price (value of shipments divided by q, deflated by the 1967 producer price index) (US Department of Commerce *Census of Manufactures*, various years)

> w = average hourly wage for production workers in wood preservatives (US Department of Commerce *Census of Manufactures*, various years)
>
> r = real user cost of capital for wood products (Wharton Econometrics, personal correspondence)
>
> h = (a proxy) lagged total revenue: real price times quantity, lagged appropriately
>
> g = government scientist months (various FPL attainment reports)

The public and private research lags are each only one year. This is shorter than in previous industries, perhaps because improved wood preservative technologies seldom require new equipment and perhaps because the FPL association with the American Wood Preservers Association is so close.[15] Therefore, information dissemination is easier and more rapid than for research breakthroughs in other forest products industries.

Table 5.10. FPL effort in wood preserving research, 1950-1980 (in 1967$)

year	scientist months	cost per scientist month*	total cost
1950	36	$ 5204.54	$ 187,363.44
1951	50	4855.78	242,789.00
1952	47	5115.39	240,423.33
1953	44	5251.19	231,052.36
1954	43	5313.59	228,484.37
1955	42	5420.57	227,663.94
1956	61	5393.55	329,006.55
1957	77	5360.01	412,720.77
1958	78	5373.36	419,122.08
1959	78	5491.13	428,308.14
1960	73	5612.64	409,722.72
1961	96	5736.45	550,697.28
1962	123	5854.13	720,057.99
1963	131	6017.04	788,232.24
1964	138	6156.62	849,613.56
1965	136	6222.45	846,253.20
1966	122	6246.15	762,030.30
1967	120	6491.89	779,026.80
1968	122	6609.92	806,410.26
1969	128	6716.05	859,654.40
1970	94	6861.94	645,002.36
1971	128	6974.01	892,673.28
1972	141	7029.79	991,200.39
1973	130	6735.11	875,564.30
1974	97	6731.16	604,422.52
1975	82	6396.18	524,486.76
1976	119	6266.84	745,753.96
1977	148	6229.98	922,037.04
1978	126	6269.30	789,931.80
1979	141	5964.67	841,018.47
1980	89	5457.57	485,723.73

* Includes overhead
Source: FPL attainment reports

In any case, FPL personnel anticipate shorter lags in this industry and our statistical tests support them.

Once again, the demand function is more problematic. The buyers of preserved wood products are diverse (railroads, telephone companies, homebuilders, farmers, users of marine pilings). These users face different wage rates and different capital costs. Inclusion of all of these independent variables in the demand equation would lead to a nightmare of collinearity.

Experiments with several alternate demand forms reveal a preference for specifications with a trend for the business cycle. The intuitive justification is that downstream consumers are so heterogeneous that, taken together, their expansions and contractions best reflect the general economy rather than any single element of it.

$$q_t = \beta_0 + \beta_1 p_t + \beta_2 b_t + \beta_3 t + \beta_4 t^2 \qquad (5.6)$$
$$\quad\;\; (-) \quad\;\; (+) \quad (?) \quad (?)$$

where b is the log of net sales in manufacturing industries (*Economic Report of the President* 1987) and t is the time variable proxy for exogenous changes in the level and use of treated lumber in downstream industries. The parenthetical expressions below the demand equation show the anticipated signs of the unknown coefficients.

Real net sales performs better than other proxies for the business cycle. The two time variables permit exponential adjustment in the industry but their signs are uncertain. Exponential decline is plausible for railroad consumption of crossties and exponential expansion is even more likely for the use of many inorganic preservatives and for recent residential construction uses of treated lumber.

Table 5.11 shows the NL2SLS estimates for both the supply and the demand equations, equations (5.5) and (5.6). All coefficients in the supply equation, except the coefficient for disembodied technical change θ, have the anticipated sign. The coefficients on labor α_1, capital α_2, and public research μ are not statistically significant. Three other supply coefficients are significant at the 10 percent level. The independent variables explain 93 percent of all variance in the quantity supplied and Durbin's h statistic indicates that serial correlation is not a problem.

The insignificant coefficient on public research is disappointing but not surprising.[16] The public research variable includes cost-reducing and both product quality and environmental research effort. We know that the latter two have little or no relationship to our measure of quantity. If (a) they dominate and (b) they are not serially collinear with cost-reducing research effort, then they are unrelated to the level of cost-reducing technical change and we must anticipate an insignificant coefficient on public research effort.

Table 5.11. NL2SLS estimates of demand and supply coefficients for wood preserving

	supply		demand
labor (α_1)	0.0366 (0.0853)	(β_0)	11.6111* (4.2514)
capital (α_2)	0.0086 (0.0101)	own price (β_1)	-1.6214* (-0.6503)
H_{t-1} (η)	0.4403** (0.2156)	business activity (β_2)	0.9286* (0.2550)
G_{t-1} (μ)	0.0186 (0.0596)	time (β_3)	-0.0558** (-0.0227)
lag (λ)	0.3222*** (0.2180)	time^2 (β_4)	0.0020* (0.0004)
time (θ)	-0.0142** (-0.0071)		
R^2	.93		.64
Durbin's h	-0.05		---
Durbin-Watson	---		2.030
degrees of freedom	25		24

Numbers in parentheses are standard errors.
  * Significant at the 1 percent level in a one-tailed test
 ** Significant at the 5 percent level in a one-tailed test
*** Significant at the 10 percent level in a one-tailed test

Of course this masks statistical confidence in our estimates of cost-reducing public research.[17]

The negative sign on θ, reflecting negative disembodied technical change, is unusual. It probably indicates that industry-wide technical change has been unable to keep pace with either industry-wide product standards or, more likely, increasing restrictions on petroleum residuals. That is, research causing decreasing final levels of residuals may not have progressed rapidly enough to maintain industry production at the old levels existing before the new higher input costs and more severe environmental restrictions. In any case, while this negative coefficient reflects unexplained relative industry decline, it has no impact on our measurement of the benefits of cost-reducing (process) research.

All demand coefficients are statistically significant at the 5 percent level or better. The positive coefficient on business activity β_2 suggests that demand is procyclical, as expected. Demand is also price elastic, as expected, because there are many substitutes for treated wood products (untreated wood, metal and concrete posts, etc.). Differentiating the antilogarithm of the demand function with respect to time shows that demand decreased through 1964 and increased thereafter, *ceteris paribus*. The lower R^2 is not surprising for this *ad hoc* specification of what should properly be a derived demand for a heterogeneous collection of consumers. The lower R^2 in this equation apparently does not suggest an absent term because absent terms cause patterns in the error terms. Yet there are no obvious patterns in the error terms of either equation. Regardless of the potential estimation problem, it causes no problems for our analysis of research benefits so long as any potentially absent term is not collinear with the price coefficient. The price coefficient (which is the price elasticity in this log linear model) is the only demand information used in estimating research benefits.

Results: The Efficiency of Public Wood Preservatives Research

Public research induces an associated private effort where specialized applications are necessary for the adoption of the new public technologies. Once more, we have no general estimate of the magnitude of this private development effort required in the wood preservatives industry. Rather, as in the softwood plywood and sawmill industries, we obtain a measure it from knowledge of a single representative case, and then test our research benefit estimates for sensitivity to variations in this measure.

Our representative case is the visual screening techniques for examining wood prior to treatment.[18] Industry implementation of these techniques began in 1968. Visual screening requires an additional employee per plant—which converts to a private expenditure of $0.12 per plant for every public research dollar. (There were 262 plants in 1950 and 498 plants in 1980.) We will compare gross public wood preservative research benefits with the sum of public research costs plus this additional induced private cost and again with a 50 percent increase in this cost to $0.18 per plant. Greater induced private development costs imply lower net economic benefits and lower rates of return to public research.

Table 5.12 displays most of our summary results for the two cases where publicly induced private development costs are $0.12 and $0.18 per plant and for the range of social discount rates between 4 and 10 percent. Net returns to producers are negative for both R&D cost alternatives and for the full range of social discount rates. Net producers' surplus is positive for, at most, six or seven individual years in the period from 1950 to 1980. This means that producers would not have conducted this research themselves. Net consumers' gains are positive for the $0.12 multiplier but negative for the $0.18 multiplier, regardless of social discount rates in our range. The combined net benefits to consumers and producers are positive in all cases. (The net benefit to society is greater than the calculated sum of consumers' plus producers' benefits by the amount of total R&D expenditures. Rows 1-3 and 8-10 in table 5.12 reflect this.) The positive

Table 5.12. Returns to public investment in wood preserving research, 1950-1980 (in millions of 1967$)

multiplier		social discount rate		
		0.04	0.07	0.10
.12	NPV^{cs}	86.1	61.9	49.1
	NPV^{ps}	-118.1	-68.3	-40.1
	NPV^{neb}	384.1	251.9	179.4
	BC^{cs}	1.16/1	1.16/1	1.17/1
	BC^{neb}	1.85/1	1.85/1	1.87/1
	IRR^{cs} not reported: multiple solutions exist			
	IRR^{neb} - 293%			
.18	NPV^{cs}	-117.3	-64.2	-34.1
	NPV^{ps}	-321.6	-194.4	-123.4
	NPV^{neb}	180.7	125.7	96.2
	BC^{cs}	0.78/1	0.78/1	0.79/1
	BC^{neb}	1.24/1	1.24/1	1.25/1
	IRR^{cs} not reported: multiple solutions exist			
	IRR^{neb} not reported: multiple solutions exist			

social gain, yet negative producer gain, justifies the public FPL presence.

Table 5.13 shows the annual sequence of net consumers' and producers' surpluses and net social gains for the single case of $0.12 private development costs and a social discount rate of 4 percent. This table shows the periodic switching from positive to negative values that prevents us from obtaining solutions for the various internal rates of return in table 5.12. Table 5.13 also provides insight to why the benefit-cost ratio increases with greater social discount rates. For example, for consumers, net losses occur in the later years and are, therefore, discounted more heavily than the larger net gains of the earlier years.

Table 5.14 reports the periodic annual values of marginal products (*VMP*), the long-run *VMP*, and the marginal internal rate of return (*MIRR*). Long-run *VMP*s less than one and *MIRR*s less than the social discount rate indicate that there was overinvestment in public wood preservatives research. This observation is all the more true for more recent years in our thirty-one year period from 1950 through 1980. These were also years of increasing petroleum product prices and the years of largest research investments in controlling environmental residuals. Removing the costs of these latter environmental and product-quality research efforts may well raise the *MIRR* above any reasonable estimate of the social discount rate and remove the question of overinvestment.

In sum, if we consider all public research investments to be of the cost-saving (process) variety, then investments in wood preservatives research were socially wise ($NPV^{NEB} > 0$) but would not have been made by private industrial investors ($NPV^{PS} < 0$). It would have been even wiser, however, for the FPL to invest at a lower total level each year ($VMP^{LR} < 1$, *MIRR* < social discount rate). The net social gains (NPV^{NEB}) would have been greater than those observed.

When we acknowledge substantial product quality improving and environmental investments in wood preservatives research, then we know that our summary benefit measures are all lower estimates. Returns to public wood preservatives research were at least as great as those we report for cost reducing-research and our subjective judgment is that the net public gains may have been positive in all scenarios.

SUMMARY AND CONCLUSIONS

Chapter 3 provided conceptual organization for our inquiry into the benefits of public research in the forest products industries. Chapter 4

Table 5.13. **Returns to wood preservatives research for individual years (in millions of 1967$);** multiplier = 0.12, discount rate = 0.04

| year | net present value of returns to: | | |
	consumers	producers	consumers & producers
1950	12.0	4.6	22.7
1951	13.4	4.7	26.1
1952	13.8	4.9	26.9
1953	12.2	4.0	24.2
1954	11.0	3.3	22.4
1955	10.7	3.0	22.0
1956	11.3	1.8	25.2
1957	10.3	- .1	25.6
1958	3.9	-4.1	15.6
1959	4.6	-1.0	17.1
1960	5.3	-3.4	18.1
1961	.4	-8.8	13.7
1962	-4.8	-14.8	9.9
1963	-5.8	-16.8	10.2
1964	-6.6	-18.7	11.1
1965	-7.1	-19.3	10.8
1966	-1.1	-14.6	18.6
1967	-1.0	-15.2	19.6
1968	- .5	-15.6	21.6
1969	-6.4	-20.3	14.0
1970	2.8	-10.9	22.8
1971	-5.7	-21.0	16.5
1972	-6.7	-23.7	18.1
1973	-1.5	-18.8	23.7
1974	14.6	- 4.1	41.8
1975	6.5	- 7.5	26.9
1976	-1.9	-17.7	21.1
1977	.7	-20.5	31.8
1978	11.4	-11.8	45.2
1979	5.5	-17.0	38.3
1980	19.7	- .3	48.9

Table 5.14. Long-run VMPs and MIRRs of public wood preserving research, 1950-1980 (1967$)

	multiplier - .12 social discount rate			multiplier - .18 social discount rate		
year	0.04	0.07	0.10	0.04	0.07	0.10
1950	1.02	0.98	0.94	0.69	0.66	0.64
1951	0.84	0.80	0.77	0.56	0.54	0.52
1952	0.86	0.83	0.80	0.58	0.56	0.54
1953	0.82	0.79	0.76	0.55	0.53	0.51
1954	0.77	0.74	0.71	0.52	0.50	0.48
1955	0.76	0.73	0.70	0.51	0.49	0.47
1956	0.58	0.56	0.54	0.39	0.37	0.36
1957	0.47	0.45	0.44	0.32	0.31	0.29
1958	0.35	0.34	0.32	0.24	0.23	0.22
1959	0.36	0.35	0.33	0.24	0.23	0.22
1960	0.38	0.37	0.35	0.26	0.25	0.24
1961	0.27	0.26	0.25	0.18	0.18	0.17
1962	0.21	0.21	0.20	0.14	0.14	0.13
1963	0.21	0.20	0.19	0.14	0.13	0.13
1964	0.20	0.20	0.19	0.14	0.13	0.13
1965	0.20	0.19	0.19	0.14	0.13	0.13
1966	0.25	0.24	0.23	0.17	0.16	0.15
1967	0.25	0.24	0.23	0.17	0.16	0.15
1968	0.25	0.24	0.23	0.17	0.16	0.16
1969	0.21	0.20	0.20	0.14	0.14	0.13
1970	0.30	0.28	0.27	0.20	0.19	0.18
1971	0.22	0.21	0.20	0.15	0.14	0.14
1972	0.21	0.21	0.20	0.14	0.14	0.13
1973	0.24	0.23	0.22	0.16	0.16	0.15
1974	0.39	0.38	0.36	0.26	0.25	0.24
1975	0.34	0.33	0.32	0.23	0.22	0.21
1976	0.24	0.23	0.23	0.16	0.16	0.15
1977	0.25	0.24	0.23	0.17	0.16	0.15
1978	0.32	0.30	0.29	0.21	0.20	0.20
1979	0.38	0.26	0.25	0.18	0.18	0.17
1980	0.46	0.44	0.42	0.31	0.29	0.28
average VMP	0.35	0.34	0.32	0.24	0.23	0.22
MIRR	< 0 (average VMP at zero discount rate - .37)			< 0 (average VMP at zero discount rate - .25)		

provided a thorough empirical application for the softwood plywood industry. This chapter provides three further applications—for the sawmill, woodpulp and wood preservatives industries. Table 5.15 provides the summary results.

The sawmill and woodpulp industries are the most recognized and probably the most important industries in the forest products sector of the US economy. Most research in these two industries is designed to decrease the costs of producing existing outputs. The wood preservatives industry represents a specialized second case. Much wood preservatives research is

Table 5.15. Returns to public research investments (1950-1980) in selected forest products industries

	softwood plywood	sawmills	woodpulp	wood preservatives
NPV (consumers' surplus)[c]	$1.32	$38.97	$.018	$.062
(producers' surplus)[c]	$1.50	-$17.56	-$.022	-$.068
(net economic benefit)[c]	$2.84	$25.96	$.004	$.252
IRR on NPV (consumers' surplus)	326%	34%	33%	multiple solutions
(net economic benefit)	499%	28%	15%	293%
VMP (short run)[a]	$12.49	- - - -	- - - -	- - - -
VMP (long run)[b]	$58.17	$2.62	$.44	$.34
MIRR	299%	16%	<0	<0

[a]In 1967$
[b]In 1967$ discounted at 7 percent
[c]In billions of 1967$ discounted at 7 percent. Consumers' and producers' surpluses are each net of total research costs. Therefore, net economic benefit = net consumers' plus net producers' suprluses plus total research costs.

designed either to extend product life or to decrease the level of environmental residuals created in the production process. These are equivalent to quality changes, rather than cost reductions, in the final output.

We want to observe (1) the statistical results of each empirical supply-demand inquiry as indicators of confidence in our final conclusions about research investments, (2) the returns on historical (1950-1980) research in the three industries, and (3) the differences in research results across the four industries, including the softwood plywood industry examined in chapter 4. What might explain the differences between industries and what do these differences suggest that might be instructive for public research managers and for the allocation of future forest product research investments?

The sawmill analysis furnishes very satisfactory statistical results—on the supply/production and the demand equations and on all coefficients in both equations. One notable term in these equations is the long (five year) research-implementation lag. This long lag is probably due to the large number, small size and diffuse nature of mills in the industry. Average returns to public research in this industry fall in the range of 17-50 percent annually for consumers and 13-49 percent for society as a whole. Producers are net losers. This means that, for producers, decreased production costs are largely passed on as lower consumer prices, causing producer revenue losses that cannot be offset by gains from increased output levels. It also means that sawmill managers have little incentive to invest in research and that the social gains from sawmill research would not be forthcoming if we relied on private industrial research. Marginal returns are between 5 and 31 percent annually. Most estimates of the social opportunity cost of capital would suggest that this is in the range of social optimality.

Statistical tests on both the woodpulp supply/production equation and on its individual terms are satisfactory. The demand equation is unconvincing. Perhaps a constant coefficient form would better explain the woodpulp consuming industry. Our analytical solution is to employ the statistically significant public research term from the acceptable woodpulp supply equation, together with an assumption of inelastic demand in our calculations of woodpulp research benefits. Inelastic demand is consistent with both prior woodpulp literature and more casual observations of the high fixed cost nature of the woodpulp-consuming paper industry. Average returns on public woodpulp research are 33 percent annually for consumers and 15 percent for society as a whole. Producer gains are negative, as they must be in an industry facing inelastic demand. Marginal returns are negative, indicating overinvestment in woodpulp research.

It turns out, however, that some woodpulp research has dealt with environmental residuals. The benefits of this research are not reflected in our measures of average and marginal returns. Therefore, these measures may be lower-bound estimates for the true average and marginal returns.

Some unproductive 1970s research featuring reduced-pollution and energy-saving pulping processes was mandated by higher government authority—contrary to the best judgment of Forest Products Laboratory scientists. This research investment would not have been made in the absence of political intervention and the average and marginal returns to woodpulp research would have been higher. This is yet another reason why our woodpulp results are lower-bound estimates for the performance of FPL scientists and their managers.

The wood preservatives analysis provides the least satisfactory statistical tests. Our demand specification is most significant but the supply/production equation is less statistically reliable. The term for private research investment is significant. The term for public research has the correct sign but is statistically insignificant—perhaps because of the mismatch between research costs which reflects process as well as product-quality research, and research benefits which only measure process technical change. We proceed to use the public coefficient anyway (as is common in the research evaluation literature where correct signs but statistical insignificance are the norm). The average social return on public wood preservatives research may be as great as 293 percent annually. (Returns to consumers and producers are indeterminate.) Marginal returns are negative, indicating either overinvestment or significant research expenditures for unmeasured product-quality improving wood preservatives research.

It should be interesting to consider the similarities and differences in research production across the four industries, including softwood plywood. The first summary observation is that public research investment was justified at some level in all four industries. That is, there were substantial net social gains to public research investment in all four. Yet

- the benefits of research would be spread too thinly among eighty establishments to justify much private research investment in softwood plywood. Research benefits would be spread even more thinly among 6800 or more sawmill firms. Indeed, we observe virtually no private research conducted in this industry.
- inelastic demands in sawmills and woodpulp indicate that most research gains in these industries are passed on to consumers. This explains why concentrated industries, like the woodpulp-consuming paper industry, might vertically integrate in order to absorb those consumer benefits.
- wood preservatives and, to a lesser extent, woodpulp producers cannot capture the benefits of those research improvements which decrease non-market environmental residuals occurring in the production process.

In sum, all four industries can justify public research at some level on basis of the resulting consumer and net social benefits. The minimal (often even negative) and intermittent producer benefits, the inability to establish lasting proprietary rights to these benefits, and the unlikelihood of sharing research costs make large private producer investments in research unlikely in these industries. These conclusions are insensitive to reasonable adjustments in both private research implementation costs and the social discount rate.

Measures of marginal returns indicate underinvestment in softwood plywood and sawmills, and overinvestment in woodpulp and wood preservatives. Smaller investments are more difficult to measure precisely. This may explain overinvestment in woodpulp. Conjectured non-measured benefits from product-quality improving research in woodpulp and wood preservatives would boost both average and marginal returns in these industries. They may be sufficient to boost these returns to the range of social efficiency.

Schumpeter asked whether large firms or firms with market power invested more and performed better. In fact, the results in these four industries suggest that neither firm size nor market power are correlated with research productivity. Average returns are greatest in softwood plywood and wood preservatives, which are the middle industries of these four with respect to firm size and market power (measured as industry concentration). Instead, the small size of the research programs in softwood plywood and in woodpulp, the long research-implementation lag in the diffuse sawmill industry, research designed to reduce environmental externalities in the wood preservatives and perhaps the woodpulp industries, and ill-conceived political encouragement of some woodpulp research all appear to be more important explanatory factors of research performance than Schumpeter's firm size or market power.

In conclusion, public research performance in these four industries, for the period 1950-1980, ranged from good to superior. It brought social gains greater than usually anticipated for private investments, comparable to public agricultural research investments (Ruttan 1980), and substantially in excess of public investments for non-industrial private forest incentives (Boyd and Hyde 1989, ch. 3) or public forest timber management (Boyd and Hyde 1989, ch. 8). This compliments research managers of the 1950-1980 period and it should encourage current research managers to sustain their best judgments.

NOTES

1. Confirmed in personal communication with J. Danielson, Forest Products Laboratory, June 15, 1989.

2. T. Haxby provided preliminary analysis and results for this section, Sawmill Research.

3. Personal communication with D. Lewis, Forest Products Laboratory, June 12, 1989. Lewis did most of the work on the BOF project in the early 1970s. Lagged total revenue is our proxy for private R&D expenditures in our other industries but its performance is unsatisfactory in the sawmill supply equation regardless of the chosen implementation lag.

4. The process for selecting a five-year lag is identical with the process used to select the two-year research-implementation lag for SWPW in chapter 4: (1) choice of the best linear 2SLS fit for the lag in the basic sawmill production function [from equation (3.3)], (2) application of this chosen lag in the general supply equation (5.1), then (3) retests of the fully specified equation (5.1) with various similar lags. A five-year lag consistently provides the best fit in both equations.

5. There is an argument that sawmill research conducted in Sweden had substantial spillover effects on American sawmills. If this spillover is statistically significant, then it would cause large error terms and an unsatisfactory coefficient of determination. Yet the coefficient of determination is an acceptable 0.67. Furthermore, if Swedish research productivity has been regular over time, then it would argue for a positive-signed term for disembodied technical change. Yet preliminary supply regressions with terms for disembodied technical change are less satisfactory in general and the coefficient on disembodied change both has the wrong sign and is statistically insignificant. Therefore, the possibility of spillovers from Swedish research does not alter confidence in our results.

6. The standard deviation of $\mu/(1 - \lambda)$ is 0.44 and the t value is 2.07.

7. Personal communication, June 12, 1989.

8. If this observation is incorrect, and there are positive economic profits in the equipment industry, then our eventual estimates of social returns are conservative.

9. The BOF technology is an example of how research has increased the minimum optimal scale of sawmill operations. Even the least-costly BOF technology ($50,000) is greater than the full cost of some of the older and smaller mills.

10. Personal communications with D. McKeever, P. Ince, and J. Minor, Forest Products Laboratory, September 7 and 9, 1988, and December 15, 1988.

11. Missing observations were obtained by first regressing R&D expenditures on price and quantity and then estimating the missing observations as a share of that year's total revenue.

12. M. Sallyards (1985) provided the primary econometric analysis for this section, Supply and Demand.

13. This section, Historical Background, relies on Bruner and Strauss (1987) and Baechler and Gjovik (1986). Bruner and Strauss also provided preliminary statistical estimates for this evaluation of wood preservatives research.

14. See footnote 7 in chapter 3, footnote 7 in chapter 4, and footnote 5 in this chapter for previous comment on disembodied technical change.

15. For examples, FPL researchers regularly serve on the AWPA staff and 20 percent of all AWPA publication since 1905 have been written by FPL researchers.

16. Most of the agriculture literature on research evaluation is satisfied with the correct sign on this coefficient. Statistical significance was our good fortune in the previous forest product industries.

17. Alternatively, we might conclude that the true coefficient is zero. If this is true, then gross benefits would be zero and net benefits would be negative. Research would have no effect on supply, therefore no effect on price. Nevertheless, producers have adopted new methods designed at the FPL and it seems implausible that these adoptions have had no effect on supply. Therefore, we prefer to accept our estimates—and the explanation that they are as accurate as can be obtained with current analytical methods.

18. Researchers at the Forest Products Laboratory confirmed the selection of this technique as representative in its requirement for industrial modification and development in each plant. (L. Gjovik, personal communication, November 1988).

Technical Change in the Southern Pine Industry

WE EXAMINED THE BENEFITS of public research in several forest products industries and found exceptional social gains regardless of the specific industry or the specialized research investments. The obvious question that now arises is whether we might expect similar large benefits from research investments in the basic growth and management of timber.

Chapter 2 anticipates, to the contrary, that timber growth and management research has been a relatively unproductive venture. The argument of chapter 2 is that, historically, forestland and the basic timber resource have been in plentiful supply relative to labor and other forms of capital. Therefore, timber prices have been relatively low—although they also have been rising through time toward a level more comparable with long-run production costs. Accordingly, the financial incentives for timber management research have been weak in comparison with the incentives for research in forest products, agriculture, and other sectors of the US economy. The next three chapters rely on empirical evidence from the southern pine industry to examine the gains from timber management research. Their most general findings support our expectations from chapter 2.

Southern pine is a shorthand nomenclature for the four major commercial species (loblolly, longleaf, shortleaf, and slash pine) that grow in the twelve southern states ranging geographically from Oklahoma to Virginia and including all states south of these two. This has probably been the most dynamic timber producing region in the US for the last eighty years. Therefore, southern pine research provides a very interesting case, and probably the best case for finding evidence contrary to our chapter 2 hypothesis of relatively low historical returns to research in timber growth and management. We might expect research impacts on southern pine to be greater than the research impacts on less economically productive timber species in the South or on other species in other less rewarding timber producing regions of the country.

Not only is this a different industry, therefore a different policy question, than we examined in the proceeding chapters, it is also a different analytical problem. The production runs of the softwood plywood industry and the various forest products industries considered in chapter 5 are brief, often less than one day. A southern pine production run (a timber rotation) may be forty years. Furthermore, many different southern pine production runs occur simultaneously (even on adjacent acres). The result is that annual data for the forest products industries capture many complete production runs, while annual inventory data from the timber management industry, on the other hand, confuse many different production runs, none of which is complete. Therefore, it is impossible to classify aggregate timber production data by discrete periodic impacts reflecting research or any other changing productive input.

Furthermore, two additional problems complicate the assessment of research impacts in the timber management industry. First, clear and direct evidence is unavailable for the physical level of certain basic inputs: labor, capital, soil nutrients, and some measure of weather. The variety of forest ownership types; farm woodlot owners, small investors, both autonomous and large vertically integrated forest product firms, public agencies; each using different inputs in different combinations, often intending to produce different outputs introduces another important difference from the agricultural research evidence. The agriculture literature assumes homogeneous ownership types. Therefore, it can better justify using cross-sectional data. Second, research cost data are even less reliable than for the forest products industries. Many more research institutions are involved and for longer periods of time. Moreover, their expenditure records are not always readily available.

Nevertheless, the policy question of research benefits in timber growth and management still confronts us—regardless of these new data problems—and we must address it. Our approach to the question begins with the measurement of technical change in the southern pine industry. We search for a measure of technical change in the industry, with the knowledge that technical change is the output of research. Comparing the benefits from technical change with a prospective range for historic research costs provides an estimate of the net research benefits in southern pine.

Our measure of technical change derives from shifts in production—in contrast with our approach in chapters 3-5 which derives technical change from shifts in the dual of the supply function. In the timber growth and management case, corrected changes in the standing softwood inventory provide a measure of the production shift, or technical change. The shift in standing inventory due to technical change causes a corresponding outward (or downward) shift in the timber supply function. Our analysis proceeds simultaneously for the two major southern pine products, solidwood (lumber and plywood) and pulpwood. Two independent econometric estimates of (1) production and (2) supply and demand form the bulk of this chapter. The final supply shifts implies changes in both producers' and consumers' surpluses for both solidwood and pulpwood stumpage. These are the gross research benefits, the results of our analysis and the feature of the final section of this chapter.

A glimpse forward to our results reveals a rate of technical change less than 1 percent annually and annual gross research benefits in the range of $0.6 million to $6.5 million. The distribution of these benefits favors consumers and, to a lesser extent, solidwood producers. Pulpwood producers are actually net losers, as research induces pulpwood price decreases. These price decreases force pulpwood producers to transfer more than their own full research gains to consumers. This provides one explanation for the continuing willingness of southern pulp and paper producers to rely heavily on non-industrial private (NIPF) timber suppliers. This also causes NIPF suppliers, rather than millowners, to suffer the burden of research-induced pulpwood price decreases.

Gross benefits are only one answer to the many questions about the research contribution to southern pine production. Chapter 7 more completely examines the distribution of these gross research benefits within the local timber producing communities or within landowner classes and among forest workers, capital inputs to timber production and higher level consumers of solidwood and pulpwood. Chapter 8 considers a range of feasible cost estimates, the resulting net social gains from public research, and various other immediate policy implications regarding wise investments in southern pine growth and management research.

CHANGES IN SOUTHERN SOFTWOOD PRODUCTIVITY: THE PRODUCTION ANALYSIS

The southern United States has witnessed a dramatic change in the composition and management of its softwood timber resource. Unmanaged old growth pine forests once covered much of the southern landscape. These forests have become third (and even fourth) growth forests in which management inputs are major factors of production. Plantations have increased and natural pine stands have continually decreased as percentages of total softwood acreage over the past thirty years. Furthermore, species that were once the major components of the ecosystem, longleaf and shortleaf pine, now compose relatively small shares of the total of southern pine inventory.

There has been an equally dramatic change in the mix of products from southern pine. Lumber was the primary product from southern pine stumpage prior to 1930. Now pulpwood stumpage consumes fully half of softwood output. Plywood and particleboard, both derived from technologies introduced to the region after 1963, are also major consumers of softwood stumpage. These increasing demands have caused total annual softwood removals from southern forests to grow from 3.1 billion cubic feet in 1952 to 4.5 billion cubic feet in 1976. (See figure 6.1.) This pattern of growth is expected to continue into the 21st century (USDA Forest Service 1982, p. 423).

Detailed insight into these dramatic changes depends on an understanding of the underlying production functions. Previous estimates of biological production functions in forestry follow either of two approaches. Many focus on the effects of site quality and stand age on the timber volume originating from a fixed area, usually one acre. More recent estimates incorporate management controlled factors such as fertilization and planting density, along with an exogenous measure of site quality, in describing productivity for a particular species from a fixed area. (Consider, for example, yield tables for managed stands of Douglas-fir or loblolly pine.)

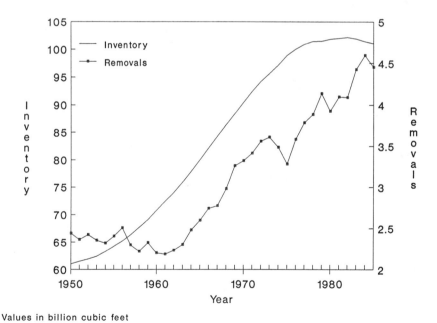

Figure 6.1. Southern softwood timber inventory and removals: 1950-1985

Most recent estimates are designed to improve understanding of the optimal combinations of timber management inputs and outputs (*e.g.*, Nautiyal and Couto 1981). Their data sources are generally cross-sectional measurements from stands of varying age and condition classes.

Measurements of aggregate technical change, however, require an understanding of historic, rather than the cross-sectional, effects of management and other inputs on timber production. They require productivity measures that reflect *aggregate* output from *all* productive acres, but measures that also abstract from productivity changes due to inputs other than research. Two recent analyses use an approach to estimating biological production that fulfills these requirements (Wallace and Silver 1984, Wallace and Newman 1986). They use county-wide time series data to assess the effects of biological characteristics (like stand density and diameter class), policy variables, ownership, and changes in forest types on total forest (hardwood and softwood) productivity.

Our analysis modifies this approach in three important ways: (1) The unit of analysis is the state, not the county. The greater aggregation in the state data decreases the degree of error attributed to small sample size in county volume and acreage data. (2) The state data are complete over a longer time period. Therefore, our analysis reflects a longer historic review, and it permits a more reliable assessment of both total and incremental changes than either the Wallace-Silver or Wallace-Newman analyses. Our earliest data originate from a 1947 South Carolina survey and our most recent data originate from a 1988 Arkansas survey. (3) Hardwood data are excluded. The relevant data are softwood acreage and production. Southern pine is more than 90 percent of the total softwood inventory the region and the major portion of forest management research in the South has to do with southern pine. Therefore, assessing southern softwood is virtually equivalent to assessing southern pine.

Our production approach is unusual in the general economics literature in that it does not explicitly define capital and labor as factors of production. Rather it relies on indices of ownership and forest management type as indicators of the relative amounts of capital and labor in production. Measurements of the biological and physical attributes of softwood timber inventories combine with the management variables to control for quality changes in forest research over time. Finally, a modified nonlinear Cobb-Douglas functional form maintains homogeneity of the production function with respect to total acreage.

Analytical Approach

Productivity is a function of both biological and land use variables.

Specifically, total production $V_{j,t}$ in state j and time period t is proportional to the total softwood forest acreage $A_{j,t}$ in that state and that period. A constant of proportionality K reflects state to state variation due to weather, biological and technological factors. The full relationship exhibits homogeneity with respect to acreage. That is, doubling the land area doubles production. Therefore, the basic production relationship is an identity with a simple multiplicative functional form

$$V_{j,t} = K(j,t)A_{j,t}; \qquad \begin{array}{l} j = 1, \ldots , 12 \\ t = 1, 2, 3, 4 \end{array} \qquad (6.1)$$

(A subsequent adjustment will permit assessment of technical change as a residual time variable.) We will examine the summary variables V, K and A in turn.

The measure of productivity $V_{j,t}$ must include both current production and production between the last and the current survey of forest inventory. Our measure is total live standing timber inventory plus total annual timber harvests and other removals during the previous ten year Forest Service survey cycle.[1]

The constant $K(j,t)$ reflects direct external inputs such as weather, soil chemistry and stand age. Expressions for these factors are not generally available in useful forms, but three biological factors; average tree diameter $B_{1j,t}$, average percent stocking or density of growing trees $B_{2j,t}$, and average site quality or growth potential of the land $B_{3j,t}$; summarize the affects of these cultural factors on the condition of the total forest stand. A Cobb-Douglas form for the relationship between these biological variables has the advantages of computational ease and clear interpretation of results. Moreover, there is no strong reason to consider a more complex form for the biological relationships where our greater concern is with estimating the effects of time and management on output. Thus

$$K(j,t) = \beta_0 \prod_{i=1}^{3} B_{ij,t}^{\beta_i} \varepsilon_{j,t}^{*} \qquad (6.2)$$

where the β_i are parameter estimates and ε^{*} represents other unspecified variables and measurement error.

The management-ownership index: This index describes total softwood acreage $A_{j,t}$ in equation (6.1). The index is

$$A_{j,t} = \sum_{k=1}^{3} \sum_{m=1}^{3} \alpha_{km} A_{km,j,t} \, \varepsilon_{j,t}^{**}, \qquad \sum_{k=1}^{3} \sum_{m=1}^{3} \alpha_{km} = 1 \qquad (6.3)$$

where the subscript k represents ownership type. The ownership categories are public land (federal, state, and other public), industry-owned or -leased land, and NIPF. The subscript m represents the management mix: pine plantations, natural pine stands, and mixed oak-pine stands. The α_{km} are parameters and ϵ^{**} is an error term. The index maintains homogeneity by forcing the unknown shares to sum to one.

The management-ownership index differentiates this model from standard production models. Management-ownership types serve as proxies for input (land, labor, and capital) variation in both qualitative and quantitative terms. Forest industry acreage, for example, reflects scale economies due to larger landholdings and better access to capital than generally available to NIPF landowners. Ownership classes also reflect different management objectives. For example, NIPF landowners may have a range of multiple use objectives while industrial landowners are more likely to have specific timber production goals. Different management objectives alter the preferred combinations of resource inputs and the rotation length and, thereby, affect total productivity on any given acre of land.

The management-ownership index is important as a control for changes in exogenous productive inputs. It may frustrate the full expression of technical change, however, because changes in ownership type, forest type and management practice, themselves, may reflect technical change. Yet controlling for them removes their shares from the final residual estimate of technical change. For example, industrial ownerships increased from nearly 20 million acres (19 percent of total softwood acreage) in 1952 to more than 28 million acres (32 percent) in 1985. (See figure 6.2.) Total pine plantation acreage increased from under 2 million to nearly 23 million acres in the same period.[2] (See figure 6.3.) The greater capital and labor inputs associated with these changes may have been justified by exogenous changes in their relative prices. They may also reflect research. That is, changes in input levels may reflect new knowledge and technical changes that show new and more profitable ways to apply the inputs. In the latter case; without research, new knowledge, and technical change; the old management-ownership-forest type configurations would remain efficient. Our eventual estimates of technical change and research benefits are both underestimates to the extent that these acreage shifts also reflect forestry research, rather than other, exogenous price-induced impacts on the management-ownership index.

Research-induced productivity shifts: Adding time period dummy variables τ_n to equation (6.1) permits measurement of the relative shifts in production over the four time periods covered by our data.

The research evaluation literature usually separates productivity shifts into movements along the production function and pure technical change (shifts in the production function) by including a direct technical change proxy along with the time dummy variables. This technical change proxy generally takes the form of either real research and extension expenditures or some measure of research activity such as publication counts. It substitutes for a measure of direct research inputs in production and thereby permits direct calculation of the shift in the production function and the marginal product of research.

Good estimates of research expenditures, however, are unavailable. Much of the research cost evidence is aggregated to the level of the two regional US Forest Service experiment stations. Therefore, the research expenditure data are not comparable with the state cross-sectional data used for our other independent variables. Furthermore, most pine research results are directly and immediately transferable across state boundaries.

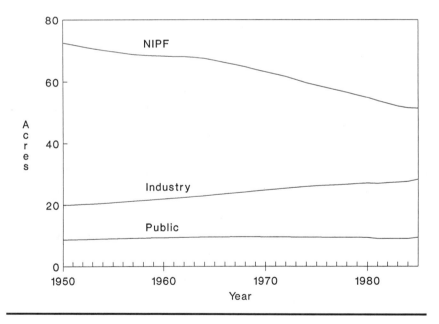

Figure 6.2. Estimated southern acreage (in millions) by forest ownership type: 1950-1985

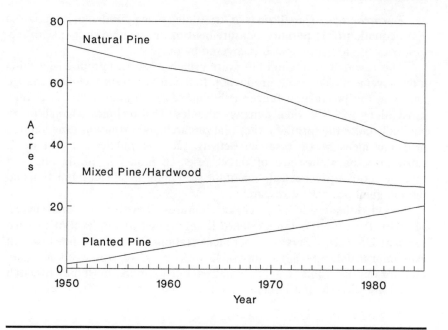

Figure 6.3. Estimated southern acreage (in millions) by forest management type: 1950-1985

Therefore, there is no justification for attempting to reallocate research costs by state. Finally, there is a long lag between the initiation of forestry research and its ultimate impact on final harvest volume. Yet, the possibility that a single research result (such as an improved seed source) may continue to have inventory impacts throughout an entire timber rotation, or even several rotations, clouds our understanding of how research alters forest inventories at any time.

Altogether, this leaves us uncertain about how research affects the production process of equations (6.1)-(6.3). Our analysis is left without a clear means to distinguish between movements along the production function and pure technical change. We rely on time dummy variables as production shifters. Our results are biased estimates of pure technical change (and research benefits) by the unknown empirical magnitude of any movement along the production function. This bias generally produces an overestimate.

The production model: Substituting equations (6.2) and (6.3) into equation (6.1) and including the time dummies yields

$$V_{j,t} = \left[\beta_0 \prod_{i=1}^{3} B_i^{\beta_i} \prod_{n=1}^{4} e^{\gamma_n \tau_n} \right] \left[\sum_{k=1}^{3} \sum_{m=1}^{3} \alpha_{km} A_{km} \right] \varepsilon \qquad (6.4)$$

The γ_n are coefficients on the time period dummies τ_n and ε is the product of $\varepsilon^*_{j,t}$ and $\varepsilon^{**}_{j,t}$. The right-hand-side of equation (6.4) drops the subscripts j and t for ease of exposition.

The model has sixteen unknown parameters but the basic homogeneity constraint ($\Sigma\Sigma\alpha_{km} = 1$) reduces the number to fifteen. Combining various management or ownership types imposes further constraints but expands the degrees of freedom in the analysis. Incorporating the constraint into equation (6.4) and performing a logarithmic transformation yields the nonlinear form ready for estimation

$$\ln V_{j,t} = \ln \beta_0 + \sum_{i=1}^{3} \beta_i \ln B_i + \sum_{n=2}^{4} \gamma_n \tau_n$$
$$+ \ln \left[A_{11} + \sum_{k=1}^{3} \sum_{m=1}^{3} \alpha_{km} (A_{km} - A_{11}) \right] + \ln \varepsilon \qquad (6.5)$$

Dropping $\gamma_1\tau_1$ in equation (6.5) avoids singularity in the estimated form and permits us to examine relative productivity changes between time periods.

We anticipate that the coefficients of the biological variables and the time dummies are positive, although of uncertain size. The coefficients of the acreage index variables should reflect the influence of increasing levels of capital, labor and technological inputs. Therefore, we anticipate positive signs and coefficient values descending from acres of plantations to natural stands to mixed oak-pine.

The relative effect of ownership on productivity is uncertain. For example, productivity may not reflect industrial owners' generally greater input use because industrial owners generally manage for shorter, financial timber, rotations. NIPF and public owners manage some stands for longer, multiple use timber rotations. Therefore, standing industrial inventories may be smaller than NIPF and public inventories. The net effect may be that the ownership index produces less distinct conclusions than the index of management type.

Data

The primary data sources are the approximately decennial forest surveys for the twelve individual states conducted by the Southern and Southeastern Forest Experiment Stations of the US Forest Service. These are the most complete records of periodic regional forest production for all ownerships. Survey procedures have evolved over the years but inventory estimates, for the most part, originate from cluster plot sampling and remeasurement of permanent plots. Data collected during each survey include standing volume, removals, mortality, evidence of stand conditions, site characteristics, ownership, and other information regarding the current status of the forest resource. Each state includes several survey regions of common geographic and forest composition. Statewide reports summarize these primary data from each survey region.[3]

Forest Survey estimates of volume per tree have changed over the years with the improvement of yield tables and utilization techniques. Therefore, we combined the measure of volume per tree by diameter class observed in the most recent of each state's surveys with the number of live softwood trees of each diameter class observed in each of that state's surveys, in order to obtain a measure for volume per acre that is consistent across survey periods. This method may understate technical change if, indeed, technical change has created higher volume trees, but this method is more accurate than introducing the spurious productivity gains resulting from more accurate growth and yield tables. Furthermore, this method avoids counting technical gains in forest products (for example, the capacity to process smaller-diameter trees) as gains in timber management.

Two variables derive from the number of trees and the average volume per tree: total live-tree softwood volume, or inventory, and the average diameter at breast height (dbh). Total volume measures are in thousands of cubic feet (Mcf) and diameter is in inches. The average diameter B_1 is a volume weighted mean value because the Forest Survey only considers trees greater than five inches dbh.

The other biological variables, average stocking B_2 and average site quality B_3, are mean values derived from the state survey reports. Stocking is the ratio of observed basal area (horizontal cross-sectional share of an acre occupied by tree stems) to the basal area of a "fully stocked" timber stand. Stocking ranges from zero to greater than 160 percent. Our summary measures of stocking are acreage weighted measures of the mean values in the state surveys. Site quality refers to the potential of an acre to produce a sustainable volume of annual growth (measured at culmination of mean annual increment). This variable is also an acreage weighted mean value.

Each state survey provides an estimate of the average volume of softwood removals during the ten year period between forest surveys. Ten year multiples of these measures provide decennial estimates. Adding decennial removals to total standing inventory at the survey date creates inventory-plus-removals, the dependent variable in equations (6.1) and (6.5).

The forest surveys summarize acreage estimates for most management-ownership type. *The South's Fourth Forest (TSFF)* revised earlier acreage estimates to identify land managed in public, NIPF, and industrial pine plantations. *TSFF* reports inventory and removal estimates for 1952, 1962, 1969, 1977, and 1985. We made no attempt to interpolate for the difference between a state's survey year and the summary regional survey year because there is little to gain from such relatively small data adjustments and the choice of summary regional survey years is somewhat arbitrary anyway.

The relatively few observations are a major weakness for our analysis. Since 1934, there have been at most five, and often only four, surveys in any of the twelve southern states. Furthermore, the manner of reporting the first survey makes those results useless for our analysis. In summary, there are 48 useful observations from the combination of twelve states and four useful survey periods.

Results

A nonlinear least-squares routine provides estimates for the parameters in equation (6.5). This section reviews the restrictive assumptions necessary to obtain the econometric results and then reviews the results themselves. The most important results for our measurement of technical change are the coefficients on the time dummy variables. The acreage indices and the distinctions between ownership and management type also invite detailed inquiry.

The limited number of observations suggests two untested assumptions regarding the parameters of equation (6.5). The first requires that production relationships between independent variables remain constant over both space and time. This is a reasonable assumption. Surely relative input levels have not changed greatly either from state to state or over the time period of analysis. The second untested assumption is that the functional form of the production relationship is the same for each of the twelve southern states. These two assumptions restrict us from assessing potentially important state-to-state production differences due to research and technical change.

Table 6.1 shows the results for seven separate regression estimates of inventory productivity. The summary regression (I) provides general insight into the determinants of regional productivity. Regressions II-VI, each based on different acreage aggregations within the management-ownership index, provide supporting evidence and additional insight.[4]

The coefficients on the three biological variables are elasticities. They show the percentage change in productivity for a 1 percent change in the biological variable. Only diameter has a coefficient greater than one—in any of the regressions. Site quality generally shows a stronger effect than stocking on productivity, although the site and stocking coefficients are not significantly different.[5]

The coefficients for the time period dummies are positive, significant, and increasing over time. These are the critical results for our analysis, the periodic inventory shifts potentially due to research and technical change. They tell us that productivity shifted upward by 7 percent from 1953 to

Table 6.1. Inventory productivity estimates

	acreage variable aggregation used						
variable	I none	II planted	III natural	IV mixed	V public	VI NIPF	VII industry
intercept (β_0)	-.348 (.600)	-.454 (.583)	-0.892 (1.470)	.070 (.440)	.072 (.393)	-.918 (.646)	-.469 (.709)
biological variables site (β_1)	.657 (.237)*	.749 (.223)*	.614 (.203)*	.513 (.180)*	.497 (.181)*	.780 (.239)*	.757 (.234)*
stocking (β_2)	.620 (.115)*	.643 (.112)*	.623 (.111)*	.630 (.114)*	.652 (.114)*	.663 (.122)*	.623 (.115)*
diameter (β_3)	1.406 (.374)*	1.286 (.357)*	1.398 (.357)*	1.350 (.355)*	1.432 (.378)*	1.518 (.382)*	1.209 (.350)*
region dummies region 2 (θ_2)	-.119 (.034)*	-.138 (.030)*	-.119 (.032)*	-.104 (.032)*	-.110 (.029)*	-.117 (.037)*	-.125 (.034)*
region 3 (θ_3)	-.168 (.055)*	-.198 (.050)*	-.157 (.047)*	-.121 (.033)*	-.128 (.035)*	-.212 (.050)*	-.187 (.055)*
time dummies period 2 (γ_2) (1953-1963)	.067 (.025)*	.062 (.022)*	.069 (.023)*	.071 (.022)*	.076 (.024)*	.030 (.021)	.059 (.024)*
period 3 (γ_3) (1953-1973)	.180 (.041)*	.183 (.036)*	.185 (.038)*	.187 (.037)*	.201 (.038)*	.127 (.029)*	.176 (.038)*
period 4 (γ_4) (1953-1983)	.195 (.050)*	.207 (.051)*	.201 (.051)*	.205 (.050)*	.221 (.052)*	.118 (.044)*	.200 (.048)*
public planted (α_{11})	-.643 (.698)	--	-2.132 (4.912)	-.785 (.575)	--	.317 (.616)	-.608 (.896)
natural (α_{12})	.643 (.316)*	.791 (.266)*	--	.353 (.148)*	--	1.059 (.719)	1.259 (.719)
mixed (α_{13})	-.890 (.522)	-1.457 (.634)*	-2.532 (3.684)	--	--	-2.584 (1.550)	-1.810 (1.163)

Table 6.1. *(continued)*

variable	acreage variable aggregation used						
	I none	II planted	III natural	IV mixed	V public	VI NIPF	VII industry
NIPF planted (α_{21})	.266 (.203)	--	1.084 (1.724)	.223 (.129)	.002 (.086)	--	.870 (.547)
natural (α_{22})	.490 (.177)*	.546 (.096)*	--	.423 (.119)*	.324 (.039)*	--	.757 (.377)*
mixed (α_{23})	.160 (.115)	.150 (.110)	.332 (.567)	--	.018 (.037)	--	.070 (.119)
industry planted (α_{31})	.633 (.284)*	--	2.637 (3.991)	.477 (.166)*	.411 (.067)*	1.056 (.686)	--
natural (α_{32})	.576 (.239)*	.618 (.195)*	--	.335 (.091)*	.431 (.076)*	1.187 (.706)	--
mixed (α_{33})	-.236 (.192)	-.092 (.137)	-.632 (1.097)	--	-.245 (.105)*	-.615 (.428)	--
mixed/natural/ *planted*		.443 (.123)*	2.249 (3.298)	-.027 (.017)			
public/NIPF/ *private*					.055 (.050)	.566 (.330)	.460 (.248)
ESS	.055	.058	.056	.059	.061	.067	.061
Error DF	30	32	32	32	32	32	32
R^2	.989	.989	.989	.989	.988	.987	.988
Restriction F[a]	--	.740	.191	.895	1.557	3.233	1.439

Total observations = 48. Parentheses show asymptotic standard errors.
* Significantly different from zero at the 5 percent level.
[a]The 5 percent $F_{2,30}$ is 3.32.

1963, by 18 percent by 1972 [or 11 percent (0.180 minus 0.067) between 1963 and 1973], and by 20 percent by 1983 [or 2 percent (0.195 minus 0.180) between 1973 and 1983].[6] There is no significant difference between the coefficients of the final two periods. (The *t* value is 1.667.) This signals a possible leveling off of productivity growth in the last decade. Nevertheless, the productivity changes that occurred by 1973 and 1983 are both significantly greater than the productivity change between 1953 and 1963. (The *t* values are 7.060 and 5.120, respectively.)

The remaining analysis of table 6.1 lends confidence to our general interpretation of the basic regression, but it does not add directly to our understanding of aggregate production. Some of the ownership information from table 6.1 will be also necessary for our assessment of the distribution of productivity gains in chapter 7.

Regional dummies: Our regressions add regional dummy variables to equation (6.5) in order to assess the presence of regional structural differences in production, as well as to measure the sufficiency of the biological variables for explaining total productivity. This "softens" the untested assumption that there are no parametric differences between states within the entire southern pine region. Differences in the regional coefficients imply regional shifts in the entire production function. They do not alter the prior assumption of no difference in the coefficients or the functional form, either between states or between regions.

Region 1 is composed of eastern Oklahoma, Arkansas, Tennessee, and Virginia; region 2 is North and South Carolina, Georgia, and Florida; and region 3 is Mississippi, Alabama, Louisiana, and eastern Texas. These regions reflect geographic and forest similarities. The productivity coefficients for regions 2 and 3 reflect significant 12 and 17 percent productivity decreases, respectively, from region 1. Regions 2 and 3 are generally considered the more important timber producing regions but site productivity and stumpage prices are generally lower in the region 1 states. These characteristics of region 1 encourage longer rotations and larger standing inventories; therefore, a larger measure of relative productivity.

Acreage indices: The interpretations of the acreage index coefficients are more complex. These coefficients measure the relative contribution to total productivity made by an acre in each management-ownership type. The relative contribution of the average acre is the inverse of the number of acreage parameters used in the model, or 0.111 when all nine management-ownership distinctions are used. For example, an industrial plantation acre contributes 53.2 percent (64.3 minus 11.1) more to productivity than the average acre. Productivity generally decreases from natural stands to plantations to mixed oak-pine stands, and (excluding mixed stands) from industrial to NIPF to public ownerships. Once more, increased inputs and better site productivity probably encourage shorter rotations and measures of natural stand productivity that generally exceed plantation productivity. Mixed stand productivity is uniformly poor in all ownerships.

Regressions II-VII: Six additional regressions more precisely display the effects of single management (regressions II-IV) or ownership (regressions V-VII) types on southern softwood productivity. For example, regression II sums plantation acreage across all ownerships but leaves natural and mixed stand classes segregated by ownership. This allows us a better understanding of the productivity effects of acreage shifts away from plantations and into the other management-ownership classes. (It also reduces the number of management-ownership classes to seven and increases the average or neutral productivity response per acre to 0.143. The management and ownership restrictions in five of these six regressions are statistically significant at the 5 percent level.)

The coefficients on the independent variables remain relatively stable regardless of the choice of regression (or acreage aggregation). The intercept term, which is not generally significant, does vary with the choice of the regression but this is due to the homogeneity assumption built into the model. The management-ownership acreage aggregation has no bearing on either the biological variables or the indicator variables for the survey periods or the regions. [Recall equation (6.1).] Therefore, coefficients on the biological, region, and period indicators in these regressions do not vary greatly so long as total acreage in the management or ownership type remains constant.

The management-ownership coefficients do vary widely across regressions Their variation suggests multicollinearity. Large simple correlations, as great as 0.85 between public and industrial plantation productivity, confirm it.[7] For this reason, we must interpret the acreage index coefficients with caution. Nevertheless these coefficients, particularly those in the final two rows of table 6.1, are instructive. The next-to-last row shows the relative contribution to productivity of plantations (II), natural (III), or mixed (IV) stands, regardless of ownership. Relative productivity descends from natural stands to plantations to mixed stands—as expected, and as previously reported for the ownerships in regression I. The last row shows the contribution of ownerships, regardless of management practice. These aggregate ownership coefficients are also supportive of our previous conclusions. Industrial and NIPF productivity is substantially greater than average public productivity.

Shifting ownership or management: A simple manipulation of the results from equation (6.5) and regression I will permit us to inquire into the productivity effects of shifting acreage from one ownership or management type to another. First differentiate equation (6.5) totally and restrict any acreage changes to be offsetting.[8] The measure for a single direct management or ownership shift is:

$$\frac{d(\ln V_j)}{d(\ln Ax_j)} = \frac{(\alpha_x - \alpha_z)Ax_j}{\Sigma^*} \tag{6.6}$$

where $\Sigma^* = \sum_{k=1}^{n} \alpha_k Ak_j;$

n = the number of acreage divisions
V_j = productivity measure
Ax_j = acreage shifted into the new management-ownership type
α_x, α_z = coefficients for management-ownership type shifted into and out of production, where all nine types are evaluated at their sample means.

The effects of broader changes, such as those that occur when one ownership expands by purchasing from several sources, are more interesting. The measure for an acreage shift originating from m different sources requires an additional restriction on the source acres. A reasonable restriction might be that the source acreage is proportional to the original management-ownership distribution of the other m types.[9] The mathematical statement of this restriction is

$$\frac{d(\ln V_j)}{d(\ln Ax_j)} = \frac{Ax_j}{\Sigma^*}\left[\alpha_x - \frac{\alpha_1 A1_j}{\Omega^*} - \ldots - \frac{\alpha_m Am_j}{\Omega^*}\right] \qquad (6.7)$$

where $\Omega^* = \sum_{s=1}^{m} As_j;$ s not equal to x.

Other parameters remain unchanged. Relative productivities continue to be evaluated at their sample means—for all acres in all management-ownership types.

Table 6.2 shows the results of shifts in management or ownership. The values in table 6.2 are South-wide comparative static productivity effects resulting from a 1 percent increase in the expanded forest management-ownership type. The squares in the table outline the more intuitively interesting changes in ownership and management.

Shifts into both plantation and natural pine management show very large positive productivity impacts, while shifts into mixed oak-pine management show large negative impacts. Shifts across various ownership types show relatively small productivity impacts. Productivity impacts are not symmetric because of differences in the initial numbers of acres in each type. NIPF is the largest ownership class. Therefore, it is not surprising that a 1 percent acreage increase in NIPF management yields larger productivity gains than 1 percent acreage increases for the other two ownership types. In general, the productivity effects of management or ownership shifts are smaller than the productivity effects of changes in the biological variables because changes in the biological characteristics of the total regional forest have generally greater scale effects on productivity than shifts in any single management or ownership type.

Summary Implications for Technical Change

Our aggregate analysis of southern softwood forest productivity features time dummies as measures of technical change and land management and ownership indices as proxies for levels of capital and labor inputs. The significantly positive effect of the time dummies on regional productivity is our primary observation. The annual productivity shift is on the order of 0.6 percent for the forty years of our analysis. It has declined in more recent years. The aggregate southern pine inventory increased from 61 million ft^3 to 102 million ft^3 in the period from 1948 to 1985. The 0.6 percent average annual shift, or 20 percent shift for the full period, means that technical change accounts for about 13.5 million of the 41 million ft^3 increase.

This 0.6 percent rate also reflects a slow turnover of the capital stock in *timber growth and management*. Comparing this value with those found for *forest products* industries provides interesting information regarding the relative productivity of southern forests. Risbrudt (1979), for example, found the average annual technical change for four SIC 24 (wood and wood products) and 26 (paper and allied products) industries to be on the order of 1.9 percent over a nineteen year period. Robinson (1975) found a 1.75 percent rate of annual technical change for the entire SIC 24 classification over a twenty-one year period. Greber and White (1982), Stier (1980), and

Table 6.2. Southern softwood inventory effects from a 1 percent acreage increase in management or ownership type

from / *into*	public	NIPF	industry	planted	natural	mixed
public	--	-0.346	-0.353	-0.353	-0.387	-0.237
NIPF	1.419	--	0.926	0.931	0.706	1.696
industry	0.596	0.422	--	0.405	0.317	0.705
planted	0.042	-0.040	-0.049	--	1.217	1.294
natural	2.663	2.247	2.203	0.948	--	2.133
mixed	0.123	0.120	-0.180	-0.178	-0.056	--

All acreage inputs are evaluated at the sample means for 48 observations.

Kendrick and Grossman (1980) each obtained similar results.

Although our analysis uses different methods and different time periods, it suggests that technical change has been less than half as rapid in southern pine—which is probably the best forest management case. This is an important observation, and it is consistent with the hypothesis set forth in chapter 2 that timber growth and management research has not been highly productive in economic terms. Of course, a convincing conclusion on this point awaits our chapter 8 comparison of the benefits associated with technical change with the costs necessary to obtain them. We cannot reject, at this stage in this research, the hypothesis that southern pine research productivity was low because research investment was also low.

The remaining summary interpretations from this section only lend confidence to our table 6.1 regressions. Therefore, these interpretations only lend indirect confidence to our conclusions regarding technical change. The effects of management and ownership type on productivity are consistent with expectations. Industrial plantations and natural stands on all ownerships show the highest relative productivity. These results are satisfying because industrial plantations generally apply higher levels of all productive inputs. They are also the greatest benefactors of research inputs. That industrial plantations show greater measured productivity gains supports the argument for a strong link between research and aggregate productivity.

Natural pine stands also show strong productivity effects. This may only reflect longer rotations on lower-valued sites.

The lower productivity of mixed oak-pine acres is also reasonable. Maintaining softwood productivity on these acres requires virtually no management inputs. Unfortunately, substantial amounts of natural pine acreage in the South have shifted into this forest type—and into pure hardwood stands. If this shift continues, then aggregate productivity in the region could decrease. Changes in other inputs and on other lands may not be able to offset it—unless demand also increases sharply.[10]

Finally, the lower overall effect of public ownership on productivity is not completely unexpected. Public plantations are managed for timber production and they do have significant positive effects on productivity. On other acres, where nontimber multiple uses of the forest may well predominate, the production effects are generally negative. The public lands are often of lower quality and they provide for genuinely conflicting demands. Therefore, it is difficult to draw general conclusions from productivity observations on these acres. It should be clear, however, that there are advantages in terms of timber productivity to confining productive activities to those public plantations that can contribute best to regional production and to placing minimal forest management effort on the other public lands.

SUPPLY AND DEMAND FOR
SOUTHERN SOFTWOOD STUMPAGE: 1950-1980

Our analysis to this point calculates an annual measure of technical change, a measure of the research impact on southern pine productivity, where productivity is the volume of standing timber inventory plus removals. In order to account for the benefits of technical change, this inventory measure must be converted into a measure of value. This requires demand and supply estimates for the two products of softwood stumpage, pulpwood and solidwood (combined plywood and lumber), with standing timber inventory as one determinant of stumpage supply. The purpose of this section of the chapter is to provide these demand and supply estimates. Our econometric technique is the three-stage least-squares (3SLS) approach for simultaneous parameter estimations of two concurrent markets. The 1950-1980 time period for the demand-supply analysis varies slightly from the 1948-1988 period of the inventory production analysis in order to accommodate a limitation in the stumpage price data.

The fundamental features of our analysis are the direct substitution of stumpage output between the pulpwood and solidwood markets and the assumption of profit maximizing behavior in the stumpage market. Profit maximizing behavior implies a derived stumpage demand and supply. A three input production technology; including capital, labor, and stumpage; summarizes stumpage demand. Stumpage supply derives from timber inventory as a proxy for production costs. This approach follows similar estimates of stumpage markets for Sweden and Finland (Brannlund *et al.* 1985a,b; Johansson and Lofgren 1985; and Kuuluvainen 1986).

A Model of Softwood Stumpage Demand and Supply

Softwood stumpage demand derives from the use of stumpage as a raw material either in the production of solidwood products (lumber and plywood) or in the production of pulp and paper products. Firms purchase stumpage and other inputs (labor and capital) and combine them in their productive processes. We assume a twice continuously differentiable production function for either solidwood or pulpwood products

$$Q_{ij,t} = q_{ij}(L_{ij,t}, K_{ij,t}, S_{ij,t}) \qquad \begin{aligned} i &= sw, pp; \\ j &= 1, \ldots, N; \\ t &= 1950, \ldots, 1980 \end{aligned} \qquad (6.8)$$

$Q_{ij,t}$ is the quantity of either solidwood (*sw*) or pulpwood (*pp*) production by firm j in period t; and $L_{ij,t}$, $K_{ij,t}$, and $S_{ij,t}$ are the quantities of labor, capital and raw material inputs which firm j uses in period t.

Solidwood and pulpwood products trade competitively in national and international markets. Therefore, their final good prices, F_{sw} and F_{pp}, respectively, are exogenous. The profit function for firm j in period t is

$$\max_{w_{i,t},\, r_{i,t},\, p_{i,t}} \quad V_{ij,t} = F_{i,t} q_{ij}(L_{ij,t},\, K_{ij,t},\, S_{ij,t}) - w_{i,t} L_{ij,t} \\ - r_{i,t} K_{ij,t} - p_{i,t} S_{ij,t} \tag{6.9}$$

where $w_{i,t}$, $r_{i,t}$ and $p_{i,t}$ are the prices of labor, capital and stumpage, respectively, for the particular industry.

The derivative of the profit function with respect to stumpage price is the firm's derived demand function. Demand is a function of the firm's output of stumpage and the prices of all productive inputs.

$$\frac{\partial V_{ij,t}}{\partial p_{i,t}} = \underset{(-)}{S_{ij}^{D}} \underset{(+)}{(p_{i,t},\, F_{i,t},\, w_{i,t},\, r_{i,t})} \tag{6.10}$$

where the signs below the independent variables represent the expected effects on stumpage demand of a positive change in the price of the input or the final good. The signs for the wage and cost of capital coefficients depend on whether stumpage is a technical complement or substitute with other inputs. The previous literature provides conflicting evidence with respect to these signs. Some find stumpage, capital, and labor to be gross substitutes in production (Stier 1980, Abt 1984, de Borger and Buongiorno 1985). Others, using similar estimation techniques, find wood and labor to be complements (Merrifield and Haynes 1983) or wood and capital to be complements (Humphrey and Moroney 1975). This confusion restricts *a priori* expectations regarding the coefficients on wages and capital costs.

Aggregate demand: If all firms in the region use the same production process and all face the same input prices, then the regional stumpage demand specification is the aggregate of the N individual firm demand functions.

$$S_{i}^{D}(p_{i,t},\, F_{i,t},\, w_{i,t},\, r_{i,t}) = \sum_{j=1}^{N} S_{ij}^{D}(p_{i,t},\, F_{i,t},\, w_{i,t},\, r_{i,t}) \tag{6.11}$$

One further adjustment to each of the conceptual demand specifications improves the estimation power of the system of equations. Lagged solidwood production ($S_{sw,t-1}$) is an additional explanatory variable in the solidwood equation. This variable incorporates two important demand characteristics not contained in most previous demand specification. First, lagged production acts as an expectations operator for further demand. That is, current demand is a function of future price expectations—which are commonly judged by recent market behavior and which we observe as last period's production. Second, lagged production also serves as a measure of the actual production capacity.[11] For both of these reasons, the expected sign of the lagged production coefficient is positive but its expected value is less than one.

For similar reasons, the sum of lagged production for both pulpwood and wood residues used in pulp production ($S_{PAR,t-1}$) is an additional explanatory variable in the pulpwood demand equation. The residues component better accounts for actual pulping capacity in the previous production period. For example, residues were 33 percent of total raw material for southern softwood pulp in 1980.[12]

Supply: The aggregate stumpage supply functions for the solidwood and pulpwood markets pose a different problem. The heterogeneous ownership and management structure of the southern forest complicates market aggregation of individual firm stumpage supply functions. Numerous factors influence the output of softwood stumpage by individual firms: multiple potential outputs (both timber and nontimber), long delay between production decisions, potential scale economies, technical change, large capital inputs, and the presence of various government market interventions (*e.g.*, capital gains taxation in the past, regeneration and timber stand improvement subsidies, technical assistance). These factors recommend hypothesizing a simplified supply function, yet a function that still accounts for the costs and returns from forest management.[13] That is our approach.

Aggregate stumpage supply for both pulpwood and solidwood products is a function of the received price for these two goods and the harvest costs. The volume of standing softwood inventory (I_t) serves as a proxy for productive capacity and an inverse proxy for harvest costs.

$$S_{i,t}^{S} = S_i^{S} (p_{sw,t}, P_{pp,t}, I_t)$$

$$\begin{array}{ccc} (+) & (-) & (+) \quad \text{solidwood} \\ (?) & (+) & (+) \quad \text{pulpwood} \end{array} \tag{6.12}$$

The expected signs for the price variables depend on whether the specification is for solidwood or pulpwood supply. These signs generally contrast because an increase in the price of one good leads the producer to shift production toward that good. Thus, for solidwood supply, the expected own price $p_{sw,t}$ coefficient is positive while the expected pulpwood price $p_{pp,t}$ coefficient is probably negative. Switching production from solidwood to pulpwood is relatively easy because timber used for solidwood is already large enough to be used for pulpwood.

The own price coefficient $p_{pp,t}$ for pulpwood supply is positive but the sign on $p_{sw,t}$ is uncertain. Uncertainty exists because firms often sell pulpwood as a by-product of solidwood timber harvests. Therefore, the possibility exists that the two outputs are net complements in pulpwood production and that there is a positive sign on the solidwood coefficient. This positive sign occurs if this complementary production effect is greater than the negative substitution effect caused by a relative increase in the solidwood price.[14]

Increases in the timber inventory have positive effects on supply in both the pulpwood and the solidwood formulation of equation (6.12). As inventory increases, scale economies arise in harvesting and the marginal harvesting costs decrease. This decreases the supply cost and increases total supply. Numerous timber supply studies use this assumption and several international comparisons provide empirical verification (*e.g.*, Gregory 1966, Styrman and Wibe 1986).

A market clearing assumption requiring the equality of demand and supply completes the system:

$$S_i^{S}(p_{sw,t}, p_{pp,t}, I_t) = S_i^{D}(p_{i,t}, F_{i,t}, w_{i,t}, r_{i,t}) \qquad (6.13)$$

Simplifying assumptions: The market model makes two important simplifying assumptions. First, prices and price expectations do not affect inventory. The effects of price expectations on total forest inventory are not well understood. Inventory changes occur relatively slowly over time so that precise incremental shifts as a result of price changes are difficult to measure and there is no compelling empirical evidence of the relationship between prices and inventory. Therefore, the inclusion of a price expectations operator on inventory in the market system would do little to decrease the potential for bias from this source.

We also assume that competitive stumpage markets exist. This assumption permits us to use real prices—rather than the marginal revenue from purchased stumpage—in the stumpage demand equations. Various degrees of monopsonistic or oligopsonistic power may exist in sub-regions of the South. They may exist because the cost of bringing wood to the mill is a major cost of stumpage production. It restricts the total distance for the competitive shipment of logs. The likelihood of bias from this assumption, however, is relatively small. The overall southern regional stumpage market is the most competitive in the US, if not the world. There are more active firms participating in the market and the southern forest industries in general, and the pulp and paper industry in particular, have shown extensive growth in production and extensive market penetration over the period of this analysis. The result has been increasing competition for stumpage and decreasing opportunity for extended imperfect market abuses.[15] An assumption of competitive stumpage markets is reasonable.

Data

The empirical specification of these functions relies on aggregate time series data for stumpage and other input prices. The time period runs from 1950 to 1980. Price variables are adjusted to the common base year of 1967 by the All Commodity Producer Price Index (U.S. Department of Labor 1985).

Stumpage quantity ($S_{sw,t}$, $S_{pp,t}$): Solidwood stumpage is the total quantity of softwood roundwood, in thousand cubic feet (Mcf), consumed in either lumber or plywood production in each of the twelve southern states. Total southern production is the aggregate of the state quantities. This aggregation does not compensate for roundwood exports to or imports from other regions since both are relatively small. Pulpwood stumpage is the quantity of softwood consumed as roundwood pulpwood for the production of pulp and paper products. The sources are US Department of Commerce annual lumber reports MA-24T, the Southern Plywood Association's compilation of annual plywood production, and the Southern Pulpwood Production Resource Bulletins from the US Forest Service's two regional experiment stations.

Stumpage price ($p_{sw,t}$, $p_{pp,t}$): We use a regionally weighted average stumpage price ($/Mcf) for each product and make no allowance for price differences between plywood and solidwood stumpage. The price difference for plywood peeler logs in the late 1970s was about 20 percent but no early price series exists for plywood stumpage prices. This introduces a relatively small bias since (1) the quantity of plywood stumpage production is small relative to lumber (only 16 percent in 1970, rising to a high of 27 percent in 1980) and (2) plywood logs can also be sold for solidwood.

Regional pulpwood prices are prices for wood delivered to the mill. We need price at the stump. Our solution begins by finding the difference between delivered and stumpage prices for private sales reported by the Louisiana forest extension service. These provide the only estimate of the difference between stumpage and delivered prices consistent with the duration of our time series. Those broad changes that have occurred in Louisiana's forests and forest industries are similar to those that have occurred in the other southern states. Therefore the Louisiana price difference should be representative.[16] Subtracting the Louisiana difference from the regional delivered price provides an estimate of the regional pulpwood stumpage prices. The source for all price series is Ulrich (1985).

Final good price ($F_{sw,t}$, $F_{pp,t}$): Ulrich's (1985) real producer price indices for lumber and wood products and for pulp and paper products provide final good price estimates. These national price indices are appropriate for both products since both trade in national markets and both compete against substitute goods from other regions. Furthermore, using national indices decreases the likelihood of simultaneity bias in the econometric estimation. This problem would occur if changes in southern production affect the national price. The degree of regional market power depends on the product. The pulp and paper industry exhibits much more market power than the solidwood industry.[17] Therefore, some potential may exist for price setting behavior in the paper products market. The presence of stiff competition from Canada and from other paper producing regions in the United States minimizes this likelihood.

Standing timber inventory (I): The Forest Surveys provide measures of the standing live softwood timber inventory for each southern state at approximately ten year intervals. The formula for our estimates of the standing inventory in the intervening years is

$$I_t = I_{t-1} + [G^* - (S_t - S^*)] \tag{6.14}$$

where G^* is the average annual net growth between survey years, S_t is actual stumpage consumption (or harvest) and S^* is average stumpage consumption between survey years. The net increment to inventory in any single year may be positive or negative regardless of the average growth rate for the survey period. The total southern inventory is the aggregate of the individual state inventories. The final inventory estimates are consistent with our inventory estimates for the technical change calculation earlier in the chapter.[18] The Southern and Southeastern Forest Experiment Stations provided this basic data.

User cost of capital ($r_{sw,t}$, $r_{pp,t}$): These values represent a composite of long- and short-term capital for each standard industrial classification (SIC) code. The series is part Moody's bond rate, part the industry's depreciation rate, part the effective investment tax credit rate for the industry and part the effective corporate tax rate for the industry. It is the same series provided by Wharton Econometrics (personal communication) and used in the softwood plywood analysis of chapter 4.

Wages ($w_{sw,t}$, $w_{pp,t}$): These values are the average hourly wages for workers by SIC code. They are nominal values derived by dividing the total wage bill by total hours worked (Ulrich 1983).

Wages and the user cost of capital data are national, since regional series of sufficient length are unavailable. These national data introduce an uncertain bias. The bias from the user cost of capital is probably small since well-functioning capital markets trade nationally. The bias from the wage series may be more serious since one arguable reason for the large forest industry movement to the South is the wage difference between that region and the rest of the country.

Econometric Estimation and Results

The estimation structure for the pulpwood and solidwood linear regression system is

$$S_{pp,t}^{S} = \alpha_{p0} + \alpha_{p1}P_{pp,t}^{*} + \alpha_{p2}I_{t} + \alpha_{p3}P_{sw,t} + \varepsilon_{1} \tag{6.15}$$

$$S_{pp,t}^{D} = \beta_{p0} + \beta_{p1}P_{pp,t}^{*} + \beta_{p2}F_{pp,t} + \beta_{p3}w_{pp,t} \\ + \beta_{p4}r_{pp,t} + \beta_{p5}S_{PAR,t-1} + \varepsilon_{2} \tag{6.16}$$

$$S_{sw,t}^{S} = \alpha_{s0} + \alpha_{s1}P_{sw,t}^{*} + \alpha_{s2}I_{t} + \alpha_{s3}P_{pp,t} + \varepsilon_{3} \tag{6.17}$$

$$S_{sw,t}^{D} = \beta_{s0} + \beta_{s1}P_{sw,t}^{*} + \beta_{sw}F_{sw,t} + \beta_{s3}w_{sw,t} \\ + \beta_{s4}r_{sw,t} + \beta_{s5}S_{sw,t-1} + \varepsilon_{4} \tag{6.18}$$

where the α_i and β_i are estimated coefficients and the ϵ_i are residuals.

Tables 6.3 and 6.4 display the regression results for the reduced form stumpage price equations and the demand and supply equations. The tables also include the elasticity estimates for the independent variables, each calculated at the means of their respective data. In general the signs and magnitudes of the elasticity measures for both systems of equations are consistent with our expectations from production theory and within reasonable bounds.

Pulpwood stumpage supply and demand: The statistical fit for the pulpwood supply equation is much better than for the pulpwood demand equation. Its R^2 and F values are greater and its coefficient of variation *(CV)* is smaller. The low values for the Durbin-Watson *(DW)* statistic in both equations indicate autocorrelation, even with the addition of lagged production $S_{PAR,t-1}$. This may be a result of the poor fit for the reduced form of the pulpwood price equation. Parameter estimates in the presence of autocorrelation are unbiased but inefficient. Therefore, the standard errors may be biased upward.

In the pulpwood supply equation the own price elasticity is positive and

Table 6.3. Three stage least squares regression results and elasticities for the pulpwood stumpage market, 1950-1980

variable	supply	demand	$P_{pp,t}$
intercept	-776,013* (181,141)	731,725 (570,621)	76.80 (101.7)
pulpwood price ($P_{pp,t}$)	4921** (19.53) [0.23]	-9092** (5127) [-0.43]	-
forest inventory (I_t)	0.02246* (0.001260) [1.20]	-	-1.334x10⁻⁷*** (8.480x10⁻⁵)
solidwood price ($P_{sw,t}$)	491.1** (239.8) [0.08]	-	-0.0184 (0.04193)
final good price ($F_{pp,t}$)	-	1800. (6327) [0.12]	0.2939 (0.8536)
wage ($w_{pp,t}$)	-	304,257* (97,565) [0.68]	22.218* (6.500)
capital cost ($r_{pp,t}$)	-	-2,341,434* (764,869) [-0.15]	-157.7* (51.42)
lagged production ($S_{PAR,t-1}$)	-	0.2265** (0.1004) [0.28]	8.774x10⁻⁶ (1.743x10⁻⁵)
df	26	24	23
R^2	0.952	0.869	0.578
DW	1.034	1.116	1.217
CV	5.3%	9.1%	10.8%
F	171.89*	31.84*	5.25*

Numbers in brackets are elasticities evaluated at the mean of the data.
Numbers in parentheses are asymptotic standard errors.
   * Significant at the 1 percent level
  ** Significant at the 5 percent level
 *** Significant at the 10 percent level

inelastic with an estimated value of 0.23. The cross-elasticity with solidwood is positive and smaller than the own price elasticity. These results are consistent with our expectations. The inventory elasticity is positive and greater than one. Its magnitude (1.20) is large, yet not unreasonable in view of strong growth in the southern pulp and paper industry over the years spanning this analysis. All independent variables are significant at the 5 percent level or better.

For the pulpwood demand equation, own price is inelastic and significant at the 10 percent level, while the final good price is positive but not significantly different from zero. A degree of substitutability exists

Table 6.4. Three stage least squares regression results and elasticities for the solidwood stumpage market, 1950-1980

variable	supply	demand	$P_{sw,t}$
intercept	888,908* (168,763)	-1,645,097* (501,887)	-473.3* (86.43)
solidwood price ($P_{sw,t}$)	3072.3* (374.1) [0.55]	-3162.4** (1423) [-0.57]	-
forest inventory (I_t)	.00648* (0.001665) [0.39]	-	-1.14×10^{-9} (-1.575×10^{-6})
pulpwood price ($P_{pp,t}$)	-11,279* (1,694) [-0.59]	-	1.771* (0.4778)
lumber price ($F_{sw,t}$)	-	21,733* (6,943) [1.72]	4.497* (0.6641)
wage ($w_{sw,t}$)	-	114,890.7** (50,727) [0.21]	-11.899 (35.80)
capital cost ($r_{sw,t}$)	-	1,024,928 (731,365) [0.08]	319.311** (128.3)
lagged production ($S_{sw,t-1}$)	-	0.7860* (0.1299) [0.78]	0.00008 (2.737×10^{-5})
df	26	24	23
R^2	0.934	0.938	0.949
DW	1.475	-	1.699
Durbin's h	-	0.821	-
CV	6.6%	6.9%	7.0%
F	122.65*	72.62*	71.52*

Numbers in brackets are elasticities evaluated at the mean of the data.
Numbers in parentheses are asymptotic standard errors.
   * Significant at the 1 percent probability level
  ** Significant at the 5 percent probability level

between stumpage and labor, since a wage increase causes an increase in stumpage demand. Stumpage and capital, on the other hand, are slight technical complements in production. Both coefficients are significant at the 1 percent level. Finally, lagged pulpwood-plus-residue input has the expected positive sign but its impact on demand is small.

The signs of the coefficients of the reduced form price equation all satisfy expectations; although the goodness of fit, as measured by the R^2, CV and F values, is not great. The overall predictive power of the pulpwood model for stumpage is only fair. Closer examination in figure 6.4b shows that the model fails to predict some recent shifts in pulpwood production. General trends, however, track very well. The cause for this may rest in the poor predictive ability of the endogenous stumpage price variable. The poor explanatory power of the final good price contributes to it. Final good price is a composite index and, as such, may be a poor indicator of the range of prices for pulp and paper products.

These pulpwood results cannot be compared directly with results from previous analyses. Most studies of southern production model pulpwood demand as predetermined, *i.e.* completely inelastic, due to either the relatively fixed demands of pulp mills, the greater importance of solidwood production, or the presence of imperfect markets (Robinson 1974, Haynes and Adams 1985).[19] Nevertheless, the low demand elasticity supports the inelastic demand findings of previous studies. The estimated supply elasticity is consistent with, though lower than, previous sawtimber supply estimates. This result is reasonable since the regression includes solidwood stumpage, a technical complement in production.

Solidwood stumpage supply and demand: The statistical fit for the entire solidwood equation system; as measured by the R^2, CV and F values; is most satisfying. The DW statistic for solidwood supply falls in the uncertain range. The presence of the lagged dependent variable in the solidwood demand equation recommends use of the Durbin h statistic. This statistic tests as a normal deviate. Therefore, we can accept the null hypothesis of no serial correlation in solidwood demand. In sum, the predictive power of the sawtimber model is much greater than the predictive power of the pulpwood model.

The significant negative sign on the pulpwood price coefficient in the supply equation implies that pulpwood is a net substitute for solidwood stumpage. This contrasts with the previous observation that solidwood is a net complement in pulpwood production. Both observations are consistent with theoretical expectations. Furthermore, the solidwood price coefficients, while still inelastic, are much larger in magnitude than the pulpwood supply coefficients. Therefore, solidwood producers must respond more than pulpwood producers to short-term changes in market variables. Finally, the standing inventory elasticity is much lower for solidwood supply than for pulpwood supply. All solidwood supply variables are significant at the 1 percent level.

The solidwood stumpage demand function also reveals major differences from the pulpwood demand function. Own price is inelastic and significant but there is a significant and elastic stumpage response to an increase in the final good price. This response does not appear in the pulpwood equations and may be a relic of the earlier structure of the lumber industry in the South. The industry prior to the mid-1960s contained many small production units with relatively low capital inputs. It could gear up or shut down quickly in response to small changes in the lumber market. The larger, more heavily capitalized and integrated firms of the past twenty years have higher-fixed costs. Therefore, they are less quick to start up or to shut down operations and the boom-bust pattern of industry production has decreased in magnitude.

Both the labor and capital cost coefficients have small positive elasticities indicating slight substitution possibilities between these two

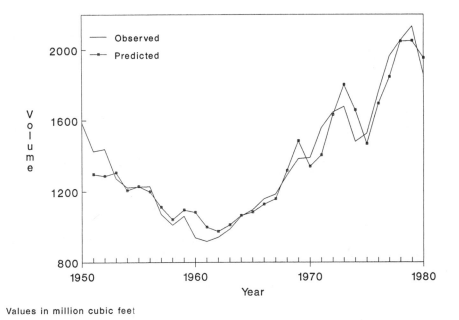

Values in million cubic feet

Figure 6.4a. Observed and predicted southern solidwood (lumber and plywood) stumpage output

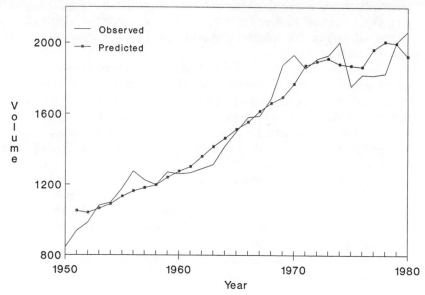

Figure 6.4b. Observed and predicted southern pulpwood stumpage output

inputs and stumpage. The substitution possibility is even less strong for capital than for labor. The capital cost coefficient is significant at the 20 percent level, while the wage coefficient is significant at the 1 percent level. The sign of the lagged production coefficient is positive, as predicted, and its value is much larger than for pulpwood. This is likely due to the altered composition of the lagged variable.

Finally, the signs of the reduced form price equations are consistent with the signs of the other equations. The effect of inventory changes on own stumpage price is much reduced from the situation in pulpwood, while the final good price has a large and significant effect.

Discussion and Implications of the Econometric Results

Our 3SLS supply and demand estimates for the southern pulpwood and solidwood stumpage markets feature supply responsive to stumpage inventory and the price of substitute outputs and demand responsive to three intermediate good productive factors, stumpage and an expectations variable. The three productive factors permit an understanding of substitution in production.

There are substantial asymmetries between the pulpwood and solidwood market structures on both the supply and demand sides. The complementary role of solidwood in pulpwood supply found in Sweden (Johansson and Lofgren 1985) also holds in the US South. Pulpwood stumpage exhibits a relatively high potential for substitution in solidwood supply. This reflects the observed behavior of southern forest landowners who have managed their timberstands on continually shorter rotations over the last thirty years.

The inventory elasticity estimate is crucial for our eventual understanding of the impacts of technical change. The pulpwood inventory elasticity is 1.20 and the solidwood inventory elasticity is 0.39. The relatively low inventory elasticity for solidwood supply, together with a general observation of longer solidwood rotations, conforms with theoretical expectations of an inverse relationship between inventory elasticity and rotation length (Binkley 1985). The inventory elasticities anticipate that the total value of expected benefits deriving from technical change, or larger inventories, will not be great because the solidwood inventory elasticity is low and the much larger pulpwood elasticity is offset by a much smaller market price.

The apparent substitution between stumpage and the other inputs, particularly labor, in the production of final goods creates a possible dilemma for policy analysts. Increasing quantities of stumpage may substitute for labor in the production process for final goods. Therefore, increasing stumpage production has a smaller effect on job creation in the processing sector and policy attempts to improve local community welfare through timber-based expansion alone may ultimately fail.

THE WELFARE IMPLICATIONS OF TECHNICAL CHANGE

The first section of this chapter assessed the impact of research investments on the southern softwood timber industry. It measured this impact as an approximate 20 percent increase in production (standing inventory plus removals) over the thirty-year period, 1953-1983. This is equivalent to average annual shifts approximating 0.6 percent, although annual productivity shifts have been smaller in more recent years. The second section estimates demand and supply for southern softwood stumpage in its two general uses, pulpwood and solidwood. It finds a pulpwood stumpage inventory elasticity of 1.20 and a solidwood stumpage inventory elasticity of 0.39.

This final section estimates the gross benefits of forest research. The inventory measure provides the link with the previous sections. Research converts into new knowledge which, in its final application, results in greater on-the-ground production, or inventory. Greater production, or inventory, expands supply and creates a new market balance. Comparing the market before and after the research impact, or inventory adjustment, yields a measure of the total social gains from southern timber research. We use the consumers' and producers' surplus approach to assess benefits of forestry research for the thirty year period from 1950 to 1980, and separate our benefit measures into shares for stumpage producers and for pulpwood and solidwood consumers.

The Consumers' and Producers' Surplus Approach

Hertford and Schmitz (1977) derive the formulae for calculating welfare effects arising from increasing stumpage supply. Their approach requires linear supply and demand equations for the stumpage market and assumes that increased production causes a parallel shift in the supply curve.[20] Their formulae for changes in the consumers' surplus CS, producers' surplus PS, and total benefit TB measures are

$$\Delta CS_t = \frac{k p_1 Q_1}{n + e}\left(1 - \frac{0.5 k n}{n + e}\right) \tag{6.19}$$

$$\Delta PS_t = k p_1 Q_1\left(1 - \frac{1}{n + e}\left[1 - \frac{0.5 k (2n + e)}{n + e}\right]\right) \tag{6.20}$$

$$\Delta TB_t = \Delta CS_t + \Delta PS_t = k p_1 Q_1\left(1 + \frac{0.5 k}{n + e}\right) \tag{6.21}$$

where k is the percentage increase in production attributed to the technological shift in production (the horizontal distance between the two supply curves divided by the value of final production Q_1); p_1 is the stumpage price after the supply shift; and n and e are the absolute values of the price elasticities of demand and supply, respectively.

The value for k is the 0.6 annual productivity change from our assessment of technical change. This estimate is conservative to the extent that some of the land area shifts to industrial ownerships and to plantation management also reflect technological responses to improved knowledge in forestry. On the other hand, the estimate is generous to the extent that it includes movements along a production function (such as those due to the changes in prices of other, non-stumpage, inputs), in addition to the shifts in the production function which reflect pure technical change.

in the production function which reflect pure technical change.

The net bias is uncertain. Therefore, we consider values for k between 0.5 and 1.0. We will even have reason in the next two chapters to inquire of the most generous estimate of 1.5 percent annual technological shifts.

The price elasticities are from the second rows of tables 6.3 and 6.4 (pulpwood supply 0.23, demand −0.43; solidwood supply 0.55, demand −0.57). Equations (6.19)-(6.21) and our general knowledge of the elasticities anticipate several important results: (1) The pulpwood supply and demand elasticities sum to less than one. Therefore, the benefits to pulpwood stumpage producers from research and technical change are negative. (2) Smaller demand elasticities are associated with larger consumer benefits. Therefore, the consumers of pulpwood (pulp and paper mills) gain more from the timber research breakthroughs than do the consumers of solidwood (sawmills and plywood producers). (3) Larger demand elasticities are associated with larger producer benefits. Therefore, solidwood producers gain more than pulp producers gain from research breakthroughs.

Welfare Results

Figure 6.5 shows the annual flow of total economic benefits accruing to both pulpwood and solidwood stumpage markets for two inventory shifts, 0.5 and 1.0 percent. The pulpwood market received greater total gains than the solidwood market during the period from the mid-1950s to 1970. This period was characterized by rapidly increasing pulpwood production with relatively high but stable prices. At the same time, solidwood production decreased and solidwood prices dropped. Subsequently, a revival in the solidwood market, with greatly increasing prices and quantities, yet stabilization in the pulpwood market, characterized the early 1970s—and the solidwood share of benefits from forestry research rebounded. Overall, pulpwood benefits are virtually equal to solidwood stumpage benefits, when both are summed without discounting, for the entire thirty year period.

Table 6.5 shows summary measures of welfare effects; including total (ΔTB), consumer (ΔCS) and producer (ΔPS) effects; for 0.5 percent and 1.0 percent annual rates of technical change in both pulpwood and solidwood markets. All values are in millions of 1967 dollars and all reflect *undiscounted* sums for the thirty year period from 1950 to 1980.[21] One initial impression is that these *undiscounted gross* values are small, two orders of magnitude less than the *discounted net* benefits of softwood plywood research over the same period. Chapter 7 will place the magnitude and distribution of these southern forestry research benefits in further perspective. Chapter 8 will examine the research costs that produced them and derive the net benefits of southern softwood forestry research.

The elasticity factors in table 6.5 show the sensitivity of these results to the elasticities from tables 6.3 and 6.4 and to double those values. In all cases, doubling reflects a greater than two standard deviation shift in each estimated own price coefficient. Linear functions and the Hertford and Schmitz assumption of parallel supply shifts insure that total benefits are insensitive to the elasticity factor. Total benefits are greatly affected, however, by the rate of technical change.

Meanwhile, table 6.5 also reflects a number of interesting distributive results. Stumpage consumers (loggers and millowners) receive approximately 90 percent of all benefits in the solidwood market. Stumpage producers (timber growers) receive small but positive benefits. Closer examination of each year in the period 1950-1980 shows these general solidwood results to be true for most specific years, although producers even receive negative benefits in occasional years. In the pulpwood market stumpage consumers receive approximately 150 percent of all benefits. This means that pulpwood producers pass to consumers more than their own full research gains. The

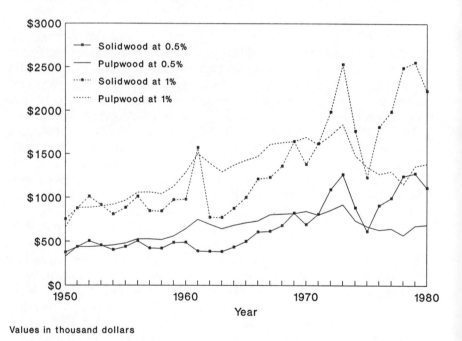

Values in thousand dollars

Figure 6.5. Combined consumer-producer surplus in the pulpwood and solidwood markets

strongly (own price) inelastic pulpwood demand and supply functions explain this phenomenon.

The intuitive explanation for these differences in the benefit distribution between the two products derives from their basic market structure. Pulp and paper production requires substantial capital outlays for plant construction but relatively small and fixed levels of input to maintain production. Therefore, increases in pulpwood stumpage supply result in lower pulpwood prices but cause only a small increase in quantity demanded. As a result, increases in timber productivity create little or no economic benefit for pulpwood stumpage producers.

For lumber production, however, the capital outlay for plant construction is relatively low and there are still a few small mills that can readily start up or close down in response to shifts both in the final good market and, to some extent, in the solidwood input market as well. This greater price responsiveness means larger increases in solidwood stumpage sold in response to productivity increases. Solidwood producers recover a greater share of their losses from selling at the lower price. This all suggests that industrial consumers of pulpwood may be happy to rely on other producers while solidwood consumers may be interested in greater corporate vertical integration. Indeed, we observe just this, with most southern pulp mills dependent on a variety of pulpwood producers. Pulpmills tend to have only small internal stumpage supplies which they reserve for internal consump-

Table 6.5. Consumer, producer, and total welfare benefits (million 1967$) in southern stumpage markets derived from shifts in southern softwood forest production, 1950-1980

elasticity factor[a]	shift = 0.5%			shift = 1%		
	consumer benefits	producer benefits	total benefits	consumer benefits	producer benefits	total benefits
solidwood market						
1x	18.44	2.10	20.54	36.86	4.25	41.11
2x	9.22	11.31	20.53	18.43	22.64	41.07
pulpwood market						
1x	31.58	-8.59	22.59	63.03	-16.86	46.18
2x	15.79	7.16	22.95	31.52	14.57	44.99

Values in this table are annual benefits summed at a zero social discount rate.

[a]Factor assumes either 1x or 2x both the estimated supply and demand elasticity for each market.

tion during periods of difficult markets.

Table 6.5 does not display the sensitivity results for elasticities only half as great as reported in tables 6.3 and 6.4 because smaller own price elasticities only increase the disparity in benefits distributed between stumpage consumers and producers. Doubling the supply and demand elasticities decreases consumer gains by 50 percent and shifts the bulk of the consumer loss to producers—for whom it is a gain. The most important effect is on pulpwood stumpage producers whose gains from research now become positive, although they remain small. Apparently our previous intuition and expectations regarding market structure holds over a wide range of stumpage price elasticities. Producer benefits are small, particularly for pulpwood. Millowners and final consumers capture the bulk of all benefits from timber growth and management research. Southern mill-owners, particularly pulp and papermill owners, are probably wise in their habit of relying on non-integrated stumpage producers.

We observe further practical understanding of these results in the support by industrial stumpage consumers for various public forestry programs, including forestry extension and incentive payments, and in land leasing and tree farming. These programs are all designed to increase production on nonindustrial private forestlands. Apparently the timber production gains from these programs largely transfer to the millowner consumers of NIPF stumpage. The equity effects may be even more disturbing. Where public research creates consumer gains but producer losses, it drives marginal producers from the market. Many marginal stumpage producers are small landowners.

Of course, even small landowners may be relatively well-off and driving them from the stumpage market may only increase the volume of southern forestland available for second homes, hunting clubs and other recreation and amenity uses of forestland. This would be an interesting and unexpected by-product of timber research.

NOTES

1. We also examined growth plus removals, measured as the ten year change in timber inventory, as an alternative measure of productivity. The eventual statistical results were similar under both the growth and the inventory formulations. The inventory measure is preferable, however, because it includes more observations and because it provides a direct tie with the inventory variable required for the supply equation later in this chapter.

2. H. Knight, R. Birdsey and R. Alig provided these preliminary estimates from *The South's Fourth Forest* (USDA Forest Service 1988).

3. Several additional sources augment data contained in the state reports. H.

Knight of the Southeastern and R. Birdsey of the Southern Forest Experiment Stations compiled additional data and personnel at these experiment stations updated preliminary acreage estimates from *The South's Fourth Forest*. Newman (1986 appendix 2.1) contains the detailed list of survey reports and tables as well as the individual state volume tables and the results of each of our data manipulations.

4. A general review of all regressions shows that their error sums of squares display relatively small differences and all regressions have high coefficients of variation (R^2). High R^2s, in part, reflect the use of *total* state *productivity*, rather than *productivity per acre*, as the dependant variable. Separate regressions using productivity per acre produce R^2s near .90 because their corrected sums of squares are substantially smaller than the corrected sums of squares in the productivity per acre regressions. Coefficients on the independent variables are relatively unchanging in the two different sets of regressions since scaling the equations [dividing both sides of equation (6.4) by $A_{j,t}$] does not alter the basic structural relationship between variables. Total productivity measures (rather than the per acre measures referred to in this footnote) are the correct basis for our analysis with its overall objective focusing on aggregate regional productivity.

5. The t value for the test for difference between the site and stocking coefficients is 0.30. The 2.5 percent test statistic for 30 degrees of freedom is 2.04.

6. Inventories were taken at different times in each state. The years associated with the periods in table 6.1 reflect the durations of the full inventory cycles. Each state inventory was completed some time during the cycle. The years identified in this paragraph; 1953, 1963, 1973, and 1983; are the midpoints in these cycles.

7. Correlations between some other similar acreage types are on the order of 0.65. Newman (1991) develops these management-ownership findings in greater detail.

8. Newman (1986, appendix 2.2) provides the supporting algebra.

9. Other distributions of source acres are also reasonable. An obvious alternative would require constant public ownership; therefore no shifts into or out of public management, and proportional shifts elsewhere. Careful understanding of the land market and production technology at the time of the shift might suggest other interesting distributions. Regardless of the specific distribution, the results of tests using equation (6.7) are useful illustrations. We anticipate that they yield close approximations to the correct estimates for the period from 1948-1988.

10. Boyce and Knight (1979) discuss this possibility.

11. The same formulation occurs if producers have an unobservable desired level of stumpage and if the adjustment toward this desired level is not an instantaneous process. Therefore, lagged production can verify either an expectations or an adjustment model.

12. The major statistical benefits from the addition of $S_{sw,t-1}$ and $S_{PAR,t-1}$ to the demand specification are a reduction in the degree of autocorrelation in the system and a large reduction in the model variance (many more significant variables, although generally with coefficients of smaller magnitude).

13. Complete specification of the production function is possible where owner specific data are available. See Brannlund *et al.* (1985b).

14. Johansson and Lofgren (1985, ch. 9) observe a case with complementary

outputs. They note that this complementarity condition implies that the present value function for the individual stumpage producer is discontinuous since the twice continuously differentiable present value function implies $\partial S_{sw}/\partial p_{pp} = \partial S_{pp}/\partial p_{sw}$. The likelihood of a discontinuous present value function as a result of discrete price changes has long been realized by forest economists. See Newman (1988).

15. For example, there were 144 pulpwood mills and 13,501 saw and plymills in the South in 1982 (USDA Forest Service 1988).

16. *Timber Mart South* (TMS, Hussey 1985) reports regional stumpage and delivered prices since 1977. The correlations of TMS regional average stumpage and delivered prices with comparable Louisiana prices since 1977 are 97 and 98 percent, respectively.

17. The South produced 65 percent of all softwood pulpwood in the United States in 1982 but only 35 percent of the softwood lumber (Ulrich 1985).

18. See Newman (1986, appendix 2.1) for the full compilation.

19. In the only other regional study of joint production in forestry, Greber and Wisdom (1985) use a mathematical programming model to explicitly examine product interdependencies between fuelwood and both pulpwood and sawlogs in the coastal plain and piedmont regions of Virginia. They systematically alter fuelwood prices and examine the impact on production behavior for the three products. They find strong complementary behavior between fuelwood and sawlogs but find that fuelwood and pulpwood are primarily substitutes. They do not examine interdependencies between pulpwood and solidwood.

20. The true form of the supply shift is unknown. If shifts are, in fact, convergent instead of parallel, then lower cost producers are able to shift production to a greater extent than marginal producers and a greater share of the total benefits accrues to stumpage producers. The same argument explains divergent shifts increasing consumer benefits (Lindner and Jarrett 1978).

21. These chapter 6 values for southern pine are *not* comparable with the measures of consumers', producers', and gross benefits calculated for the forest products industries in chapters 3-5. The southern pine values are generous estimates of undiscounted 1980$ benefits. The forest products industries values were directly estimated and statistically reliable, discounted 1967$ benefits.

Distributive Effects of Technical Change in Southern Softwood Forestry

ONE JUSTIFICATION for any public investment can be its distributive impacts. This justification usually refers to the investment's beneficial effects on a less-privileged population or a less-developed region. The purpose of this chapter is to examine the distributive justification for southern pine research. The previous chapter reported the summary distributive impacts on all producers and all consumers of southern pine. This chapter examines finer distinctions in benefit distribution.

The gross impacts of southern pine research are small, an undiscounted $40-85 million (1967$) for the sum of twelve states and accumulated over the full period from 1950 to 1980. This amounts to an annual average of only $110-130,000 for each state, a small number by any standard.

Nevertheless, the value of roundwood ranks among the first three agricultural crop values in every southern state. Roundwood is the basic input for all production from the forest products industries. The wage bill for these industries ranks either first or second among all manufacturing industries in each of the twelve southern states (USDA Forest Service 1988). Furthermore, forestry and forest products are generally rural activities and the rural South is a poor region. Therefore, it may be useful to consider whether even a small gross research impact is important for selective communities, selective industries within the forestry sector, or specialized factors of production.

The annual research-induced timber productivity increase of 0.6 percent reported in chapter 6 is our point of departure. Combining this datum with local input-output (I-O) models will allow us to estimate the sum of direct, indirect and induced impacts on local economies. The impacts on wages and employment will be most interesting because improving the conditions of unemployed and low wage workers can be one distributive justification for publicly supported forestry research activities.

We will use IMPLAN (IMpact analysis for PLANning), an I-O model developed by the US Forest Service, to examine three regional economies within the South. The first two are south Georgia and the region surrounding Bogalusa, Louisiana, two of the most active timber markets in the South. The third is coastal North Carolina. Drainage technology permits the development of growing pulp and paper and solidwood products industries along the North Carolina coast. The distinctions between these three sub-regions should permit us to observe a range of responses to research-induced increases in southern pine productivity. This range can serve to focus discussion about the likely South-wide beneficiaries of pine research.

The first section of this chapter provides explanation and justification for our approach to this distributive problem. It includes comment on I-O experience in the forestry literature as well as a basic discussion of the IMPLAN model. Subsequent sections discuss the south Georgia, Bogalusa and coastal Carolina cases in turn. Each case begins with a comment on the regional forestry and forest products industries, continues with the specialized regional assumptions for IMPLAN, and then reports the regional impacts of research-induced increases in southern pine productivity.

APPROACH

A 1950 base and a 20 percent forest productivity change over thirty years would be natural starting points for our distributive analysis. They also would yield meaningless input-output results. Thirty years is too long to hold technology constant in all other sectors of the economy. Both the production demands due to technical change in the other sectors and the interactive effects of technical change in the other sectors on forestry sector technologies might alter our distributive conclusions.

Our alternative is to seek a distributive impression from some shorter period during which the assumptions of (a) constant technology in other sectors and (b) no interactive technical changes are not disruptive for meaningful forestry sector conclusions. This approach will not yield results that are strictly comparable with results in other chapters of this book, but it will identify the forest industries and the factors of production which are significantly affected by research-induced technical change in forestry. The latter is the more important objective.

The 3-5 year period of a single macroeconomic expansion is short enough to fit this description, yet long enough to permit a significant injection of new forest productivity. We will use a 3.0 percent forest productivity change (a generous replacement for 0.6 percent per year for 3-5 years). Our base year is the 1982 base year of the IMPLAN model.

Chapter 6 found that stumpage demand is inelastic, particularly for the pulp industry. Inelastic demand insures that most productivity benefits accrue to consumers, not producers. It also justifies our focus on the forward impacts of forestry research. Bengston and Gregersen (1986) confirm this conclusion. Observations of oligopsonistic stumpage markets reinforce it.

Furthermore, even where producer gains do occur, they fail to justify distributive arguments for public support of forestry research. Distributive arguments for public support require that recipients are poor or otherwise disadvantaged. There are four factors of timber production: land, capital, managerial inputs, and labor. (Opportunity costs on land and capital over time draw the largest factor shares.) Landowners capture the returns on the first three. Landowners themselves fall into three classes: public, industrial and non-industrial private (NIPF). No one uses distributive arguments for public support of the first two. NIPF landowners may not be great industrial capitalists but, as a group, their incomes are well above national and regional averages (Boyd and Hyde 1989, ch. 3). Therefore, there can be little distributive justification for public programs supporting them either. Labor, the remaining factor, might justify public support, but its factor share in timber production is very small. Therefore, only a small share of the producer gains from southern pine research ever get passed back to laborers working in timber growth and management.

Altogether, inelastic stumpage demand, potentially oligopsonistic stumpage markets, and the lack of distributive justification for public support of the important factors of timber production, deny the need to inquire into the producer benefits of southern pine research. Our attention will be on consumer benefits and forward linkages.

I-O Modeling in Forestry

I-O models display the direct, indirect, and induced effects on an entire economy originating from injections into any particular sector in that economy. They have been used in forestry to display the importance of the sector (industry size, employment and wage effects, and linkages with other sectors), forestry's interregional impacts, and the impacts of forest policies. [See, for example, Elrod *et al.* (1972), Porterfield *et al.* (1978), Flick *et al.* (1980), Troutman and Breshears (1981), and Jones and Zinn (1986) on sector importance; Teeter *et al.* (1989) on interregional impacts; or Schallau *et al.* (1969) and Connaughton and McKillop (1979) on forest policies.]

Multipliers are the summary I-O measures of impacts on output, income and employment. Multipliers measure total impacts divided by the direct impacts.[1] They show the total impact on the regional economy of a one dollar direct injection of demand into a specific sector. The forestry literature reports extremely high output and income multipliers, averaging 2.25 for the South. Employment multipliers for the South average 2.07 for the wood products industries and 2.66 for the pulp and paper industry (Flick 1986). Large multipliers imply potentially large distributive impacts in our analyses of south Georgia, Bogalusa, and coastal Carolina.

The US Forest Service developed its own I-O model in response to a National Forest Management Act (1976) requirement for economic analyses of all national forest plans. This model, IMPLAN, is now an efficient software package (Alward and Palmer 1983, Palmer *et al.* 1985, Alward 1987). It is as well-designed as any for our purposes and it has the additional advantage of being well-known to forest policy analysts.

IMPLAN's data base consists of a national-level technology matrix and county-level estimates of activity for 528 economic sectors. IMPLAN's technology matrix comes from the 1977 Bureau of the Census national I-O model, updated to correspond to the 1982 National Income and Products Account. IMPLAN's measures of county-level sectoral activities originate with various Census sources. The resulting model loses accuracy due to its general application of national technology coefficients for county-level analysis, but this generalization also permits the important advantage of easy assembly of any number of counties for a specific regional analysis.[2]

Three final steps precede the actual calculations. The first identifies the boundaries of analysis, the counties in each case study. The second aggregates sectors of no particular importance to our analysis, thereby reducing the total number of sectors from 528 to a number appropriate to our particular problem. We aggregate the non-forestry sectors but maintain the identity of the six forest product industries: independent logging, sawmills, hardwood lumber, plywood, other wood products, and pulp and paper. The result is seventeen internal sectors, including the six in forest products, an external sector, and household industries.

Final processing sector demand drives the IMPLAN model. This feature directs the third preparatory step. We use the general assumption that the entire 3.0 percent research-induced productivity gain is fully processed within the region. (This implies that our regional impacts include all the impacts of stumpage productivity, whether the stumpage is consumed in the region or exported to adjacent regions for processing.) Processing adds value to the regional stumpage productivity estimates. Therefore, the actual exogenous demand injections to IMPLAN must reflect this value added. The pulp and paper industry absorbs the direct IMPLAN multiplier impacts of southern pine pulpwood productivity increases. The sawmill and plywood industries absorb the direct impacts of solidwood productivity increases. Sensitivity analyses can check for the allocation of total southern pine productivity between the pulp and paper and the sawmill/plywood industries. Increasing demand in these industries generates responses throughout the regional economy, including the responses in the basic timber harvesting activity corresponding to the basic 3.0 percent southern pine productivity increase. Type III multipliers, measuring indirect and induced effects, include these additional economy-wide impacts.

FORESTRY IN SOUTH GEORGIA

The 57 counties in south Georgia contain 9.8 million acres of forestland. Thirty-two percent of these are industrial, 64 percent are NIPF. The 1982 forest survey reports that 77 percent of industrial forests are pine plantations. The incentives of the Conservation Reserve Program support more recent increases in NIPF plantations and the 1988 forest survey reports that 40 percent of NIPF forests are in pine plantations. The region supports 360 million cubic feet (MMcf) of annual softwood removals, 37 percent of these are industrial and 62 percent are NIPF.

South Georgia is a hub of the southern pulp and paper industry. There are ten major pulp and paper plants along the Georgia coast, ten more on the interior coastal plain, and several more within easy pulpwood hauling distance in north Florida. There are also more than one hundred solidwood processing operations within the region.

Table 7.1 reports economic activity in the region by the seventeen I-O sectors. The six forest products industries account for 12 percent of total industrial output (TIO), 8 percent of value added, 5 percent of employment and 9 percent of wages. The combined forest products sector is the largest non-manufacturing sector. Pulp and paper is the dominant industry within the sector. It accounts for 68 percent of forestry sector TIO. It also pays the highest average wages of any industry in the region.

Table 7.2 shows the IMPLAN Type I (direct and indirect) and Type III (direct, indirect and induced) multipliers for the 57 county region. Excluding the very small hardwood lumber industry, the Type III employment multipliers for the forest products industries range from 1.80 to 2.34. They are greater than the comparable output and income multipliers. The forest products employment multipliers are also greater than the employment multipliers for all non-forestry sectors except construction and food products. The forest products income multipliers are greater than the income multipliers for all but the agriculture and food products sectors.[3]

The large forestry sector income and employment multipliers in table 7.2 indicate relatively large impacts for a given exogenous injection to the economy. Nevertheless, forestry's small share of total regional employment in table 7.1 suggests that even a large impact on the sector will not have a large impact on the full south Georgia economy. Furthermore, forestry's larger than proportional wage share suggests that forestry sector employees

Table 7.1. South Georgia sector data and selected share levels

sector	million 1982$							# jobs
	intermediate demand	final demand	TIO	wages	property income	total income	value added	employ-ment
logging	293.23	13.05	306.28	60.65	29.60	90.25	91.40	3,306.40
sawmills	125.68	135.06	260.74	72.96	16.49	89.45	93.09	3,197.51
hardwood lumber	2.42	0.48	2.90	1.03	0.10	1.13	1.18	89.93
plywood	61.74	49.44	111.18	29.02	6.90	35.92	37.35	1,461.19
other wood products	24.02	122.87	146.89	33.96	8.10	42.06	44.15	2,572.26
pulp & paper	57.35	1,686.65	1,744.00	400.49	126.92	527.42	561.18	11,021.99
agriculture	675.79	1,182.36	1,858.15	163.60	613.29	776.89	811.94	29,671.42
mining	5.83	52.36	58.19	16.03	9.37	25.40	27.17	626.17
construction	315.22	1,151.13	1,466.35	481.95	50.27	532.22	548.92	18,239.49
manufacturing	514.90	2,711.58	3,226.48	866.24	169.83	1,036.07	1,079.82	31,460.23
food products	362.51	1,093.25	1,455.76	173.50	95.66	269.16	275.69	12,821.24
textiles	208.89	1,461.09	1,669.98	416.16	95.34	511.50	521.93	29,713.12
utilities	719.38	977.17	1,696.55	446.44	294.94	741.38	806.81	18,023.94
wholesale & retail	79.79	935.64	1,015.43	468.04	114.21	582.24	731.69	59,460.16
finance	342.03	1,255.28	1,597.32	282.19	684.51	966.70	1,159.36	15,296.92
services	500.87	2,108.54	2,609.41	1,003.37	354.01	1,357.38	1,402.12	78,331.77
government	84.23	1,513.39	1,597.62	1,449.04	19.29	1,468.33	1,468.33	99,145.62
rest of world	0.00	178.91	178.91	(0.33)	179.24	178.91	178.91	(93.25)
household industry	0.00	51.54	51.54	51.54	0.00	51.54	51.54	6,019.11
inventory valuation	0.00	0.00	(39.69)	0.00	(39.69)	(39.69)	(39.69)	0.00
total	4,373.85	16,679.80	21,013.96	6,415.87	2,828.39	9,244.25	9,852.89	420,365.20
wood products %	12.9%	12.0%	12.2%	9.3%	6.7%	8.5%	8.4%	5.2%
agriculture %	15.5%	7.1%	8.8%	2.5%	21.7%	8.4%	8.2%	7.1%
textiles %	4.8%	8.8%	7.9%	6.5%	3.4%	5.5%	5.3%	7.1%
manufacturing %	11.8%	16.3%	15.4%	13.5%	6.0%	11.2%	11.0%	7.5%
construction %	7.2%	6.9%	7.0%	7.5%	1.8%	5.8%	5.6%	4.3%
services %	11.5%	12.6%	12.4%	15.6%	12.5%	14.7%	14.2%	18.6%
finance %	7.8%	7.5%	7.6%	4.4%	24.2%	10.5%	11.8%	3.6%

population = 1,189,700

Table 7.2. South Georgia Type I and III output, personal income, and employment multipliers

sector	Type I			Type III		
	output	income	employ-ment	output	income	employ-ment
logging	1.56	1.42	1.82	1.78	1.76	2.33
sawmills	1.72	1.50	1.76	1.96	1.77	2.26
hardwood lumber	1.53	1.35	1.24	1.96	1.73	1.59
plywood	1.61	1.52	1.62	1.85	1.81	2.07
other wood products	1.53	1.56	1.41	1.81	1.94	1.80
pulp & paper	1.37	1.40	1.82	1.50	1.58	2.34
agriculture	1.37	1.76	1.34	1.61	2.62	1.72
mining	1.20	1.22	1.32	1.36	1.40	1.69
construction	1.27	1.25	1.46	1.48	1.45	1.88
manufacturing	1.20	1.22	1.31	1.35	1.39	1.69
food products	1.62	1.77	1.95	1.82	2.28	2.50
textiles	1.25	1.26	1.24	1.50	1.57	1.59
utilities	1.24	1.27	1.34	1.40	1.46	1.72
wholesale & retail	1.16	1.11	1.05	1.86	1.58	1.35
finance	1.18	1.30	1.31	1.32	1.55	1.68
services	1.24	1.17	1.14	1.62	1.48	1.46
government	1.06	1.02	1.02	1.77	1.27	1.30
rest of world	1.00	1.00	1.00	0.99	1.99	1.28
household industry	1.00	1.00	1.00	2.32	1.41	1.28
inventory valuation	1.00	0.00	0.00	1.00	0.00	0.00

are better paid than the average worker throughout the region. Therefore, forestry sector employees in this region are probably not the workers for whom public agencies usually design redistributive programs. In sum, we anticipate that the distributive justification for public investment in southern pine research is probably not great.

IMPLAN Modifications

Two adjustments specific to south Georgia remain before we can use IMPLAN and report its conclusion.

A 3.0 percent increase in the 1982 south Georgia pine harvest level of 360 MMcf is approximately 40 million board feet (MMbf). We will examine the alternative allocations of (a) 30 MMbf to the pulp and paper industry and 10 MMbf to the sawmill and plywood industries and (b) 20 MMbf each to pulp and paper and to sawmill/plywood. The former allocation is a fair reflection of the South-wide allocation observed in chapter 6. It is also consistent with the strong pulp and paper industry in south Georgia. Tables 7.1 and 7.2 suggest that the latter allocation will produce more optimistic employment and wage effects. Therefore, this allocation provides an upper bound for distributive justifications for southern pine research investments.

Timber-Mart South prices (fob, 2d quarter 1982) and the value added estimates used in US Forest Service applications of IMPLAN create our estimates for per unit raw material expenditures by the pulp and paper industry and the sawmill/plywood industries. The pulpwood price is $164/Mbf (converted from state chip prices using a factor of 2.5 T/Mbf) and the value added is $0.67 for every dollar of raw material. Therefore, the final demand injection for pulp and paper is $270/Mbf. The solidwood price is $217/Mbf and the value added is $0.55 for every dollar of raw material. The final demand injection for the sawmill/plywood industries is $336/Mbf.

Results

Table 7.3 reports the results from increasing pulpwood production by 30 MMbf and solidwood production by 10 MMbf. Three forest products industries show the largest impacts but the finance, service and whole-sale/retail sectors show important indirect impacts. The pulp and paper industry gains little because the raw material factor share is so small in this industry. Property absorbs most of the impact on the finance sector. Wages absorb most of the impact on the service and wholesale/retail sectors. Agriculture also gains employees, but the largest agricultural impact is on property value, not on wages. Total employment increases by 330 persons in all sectors—on a regional base of 420,365. The forestry sector adds 183

Table 7.3. South Georgia sectoral impacts from increasing pulpwood output by 30 million board feet and sawtimber output by 20 million board feet

| sector | million 1982$ | | | | | | # jobs |
	final demand	TIO	wages	property income	total income	value added	employ-ment
logging	.000	3.024	.599	.292	.891	.902	32.64
sawmills	3.360	3.979	1.113	.252	1.365	1.421	48.80
hardwood lumber	.000	.009	.003	.000	.003	.004	.28
plywood	3.360	3.631	.948	.225	1.173	1.220	47.72
other wood products	.014	.091	.021	.005	.026	.027	1.59
pulp & paper	8.101	8.210	1.885	.597	2.483	2.642	51.89
agriculture	.041	1.231	.108	.406	.515	.538	19.66
mining	.000	.019	.005	.003	.008	.009	.21
construction	.000	.339	.111	.012	.123	.127	4.22
manufacturing	.084	.583	.156	.031	.187	.195	5.69
food products	.216	.357	.043	.023	.066	.068	3.15
textiles	.093	.137	.034	.008	.042	.043	2.45
utilities	.282	1.225	.322	.213	.535	.583	13.02
wholesale & retail	.511	.559	.257	.063	.320	.402	32.71
finance	.719	.984	.174	.422	.596	.714	9.43
services	.944	1.401	.539	.190	.729	.753	42.06
government	.075	.183	.166	.002	.168	.168	11.33
rest of world	.000	.000	.000	.000	.000	.000	.00
household industry	.028	.028	.028	.000	.028	.028	3.32
inventory valuation	.000	.000	.000	.000	.000	.000	.00
total	17.829	25.993	6.515	2.745	9.260	9.844	330.16

population change = 934

employees, or 55 percent of the small total increase.

Table 7.4 reports the more optimistic impacts from increasing pulpwood and solidwood production by 20 MMbf each. The larger employment and wage impacts are a function of the shift away from the relatively capital intensive (labor and raw material extensive) pulp and paper industry. Final demand and TIO increase over the previous scenario. Three primary forest products industries again show the largest impacts.

Comparing table 7.4 with the base case in table 7.1 shows that total employment in all sectors increases by 487, or 0.12 percent. Forestry sector employment increases by 311, 64 percent of the total employment increase, but only a 1.4 percent increase on base case forestry employment. Wage effects are sharpest in three forest products industries and unimportant elsewhere. They range from increases of 0.3 percent in pulp and paper to 1.6 percent in logging, 2.9 percent in sawmills, and 6.5 percent in plywood.

In sum, the important distributive impacts in south Georgia are not great under any scenario. Publicly funded forestry research creates minor wage and employment gains, mostly in three forest products industries. Furthermore, reflections on table 7.1 show that the small wage gains are concentrated on the logging, sawmill, and plywood industries. These are among neither the eight lowest paid nor the ten largest employing sectors

Table 7.4. South Georgia sectoral impacts from increasing both pulpwood and sawtimber output by 20 million board feet

sector	final demand	TIO	wages	property income	total income	value added	employ- ment
	million 1982$						# jobs
logging	.000	4.942	.979	.478	1.456	1.475	53.35
sawmills	6.720	7.599	2.126	.481	2.607	2.713	93.19
hardwood lumber	.000	.017	.006	.001	.007	.007	.54
plywood	6.720	7.240	1.890	.449	2.339	2.432	95.16
other wood products	.021	.152	.035	.008	.044	.046	2.67
pulp & paper	5.402	5.537	1.272	.403	1.675	1.782	34.99
agriculture	.061	1.984	.175	.655	.830	.867	31.68
mining	.000	.013	.004	.002	.006	.006	.14
construction	.000	.389	.128	.013	.141	.146	4.84
manufacturing	.123	.730	.196	.038	.234	.244	7.12
food products	.318	.516	.062	.034	.096	.098	4.55
textiles	.138	.189	.047	.011	.058	.059	3.37
utilities	.416	1.440	.379	.250	.629	.685	15.30
wholesale & retail	.755	.819	.378	.092	.470	.591	47.98
finance	1.060	1.439	.254	.617	.871	1.044	13.78
services	1.393	1.987	.764	.270	1.034	1.068	59.66
government	.110	.226	.205	.003	.208	.208	14.04
rest of world	.000	.000	.000	.000	.000	.000	.00
household industry	.042	.042	.042	.000	.042	.042	4.90
inventory valuation	.000	.000	.000	.000	.000	.000	.00
total	23.280	35.264	8.941	3.805	12.745	13.512	487.25

population change — 1,379

in the region. Therefore, the distributive merit of southern pine research is doubtful for south Georgia.

BOGALUSA, LOUISIANA AND ENVIRONS

The Bogalusa area has a 75-year history of continuous forest management. The Great Southern Lumber Company in Bogalusa maintained the largest sawmill capacity in the South from early in this century. More importantly for our analysis, Great Southern collaborated with the US Forest Service and other forestry researchers in developing regeneration techniques for longleaf and other southern pines and in applying sustained yield management techniques.

The 15 parishes and counties in Louisiana and Mississippi that surround Bogalusa contain 3.5 million acres of forestland.[4] The most recent forest surveys (1984 for Louisiana and 1987 for Mississippi) report an ownership distribution of 71 percent NIPF, 27 percent industrial, and 2 percent public. Plantation acreage and the proportion in pine have both increased since the previous survey. Annual softwood removals from the area are 108 MMcf.

The region supports a single pulp and paper facility owned by Gaylord Container Corp. and a number of solidwood plants. The adjacent regions also support a number of wood processing facilities that periodically draw on the Bogalusa region's forest inventory.

Table 7.5 reports economic activity by I-O sector. The combined forest products industries accounting for 10.8 percent of TIO, 6.9 percent of value added, 5.3 percent of employment, and 8.6 percent of wages. Pulp and paper accounts for 53 percent of the entire sector's TIO but only 42 percent of its employment. Average forestry and forest products wages are moderate for the region—except for the logging and pulp and paper industries whose workers are among the region's very best paid. Once more, forestry's small share of regional employment but larger regional wage share anticipate that forestry sector investments will not have important distributive impacts.

Table 7.6 shows the IMPLAN Type I and Type III multipliers for the fifteen parish/county region. They are similar to those for south Georgia and all multipliers fall within the range of state averages reported by Flick. The forestry sector multipliers are insufficiently large to alter our expectation of unimportant distributive effects in this region.

Table 7.5. Bogalusa region sector and selected share levels

sector	intermediate demand	final demand	TIO	wages	property income	total income	value added	# jobs employment
			million 1982$					
logging	112.11	4.35	154.95	30.68	14.98	45.66	46.24	909.73
sawmills	45.79	3.54	76.86	21.51	4.86	26.37	27.44	1,256.69
hardwood lumber	1.45	0.03	1.53	0.51	0.05	0.55	0.58	67.23
plywood	25.17	0.62	36.05	9.14	2.23	11.37	11.84	669.19
other wood products	8.77	25.92	72.05	20.26	4.27	24.53	25.32	1,479.51
pulp & paper	26.43	156.34	525.25	121.70	38.83	160.52	171.19	4,059.67
agriculture	185.4	50.64	448.35	38.75	84.67	123.43	130.93	7,533.57
mining	210.22	7.58	263.78	24.38	109.42	133.80	162.36	620.32
construction	163.04	1,030.74	1,572.59	499.33	68.91	568.24	584.91	26,027.31
manufacturing	1,638.02	1,044.21	8,792.23	827.77	404.36	1,232.13	1,429.59	21,553.28
food products	77.65	214.36	537.23	80.25	44.16	124.41	127.17	3,579.74
textiles	17.26	61.39	127.18	33.57	8.74	42.30	42.89	3,922.40
utilities	741.57	301.28	1,079.98	282.05	223.80	505.85	546.80	10,494.09
wholesale & retail	99.79	849.00	1,029.91	474.31	115.98	590.29	741.67	50,969.01
finance	563.25	1,266.39	1,970.14	270.70	903.34	1,174.04	1,421.53	14,252.40
services	837.73	1,434.36	4,177.40	1,612.88	636.15	2,249.04	2,293.27	92,269.71
government	67.47	1,410.25	1,484.24	1,396.71	13.65	1,410.36	1,410.36	76,371.49
rest of world	0.00	166.75	166.75	(0.31)	167.06	166.75	166.75	(86.92)
household industry	0.00	45.20	45.20	45.20	0.00	45.20	45.20	5,469.98
inventory valuation	0.00	0.00	(36.99)	0.00	(36.99)	(36.99)	(36.99)	0.00
total	4,821.13	8,072.94	22,524.67	5,789.38	2,808.46	8,597.84	9,349.04	321,418.40
wood products %	2.2%	2.3%	3.2%	3.0%	1.8%	2.6%	2.5%	2.3%
agriculture %	3.8%	0.6%	2.0%	0.7%	3.0%	1.4%	1.4%	2.3%
textiles %	0.4%	0.8%	0.6%	0.6%	0.3%	0.5%	0.5%	1.2%
manufacturing %	34.0%	12.9%	39.0%	14.3%	14.4%	14.3%	15.3%	6.7%
construction %	3.4%	12.8%	7.0%	8.6%	2.5%	6.6%	6.3%	8.1%
services %	17.4%	17.8%	18.5%	27.9%	22.7%	26.2%	24.5%	28.7%
finance %	11.7%	15.7%	8.7%	4.7%	32.2%	13.7%	15.2%	4.4%

total population = 955,200

IMPLAN Modifications

A 3.0 percent increase in 1984 harvest levels is 20 MMbf. We will
examine the alternative allocations of (a) 15 MMbf to pulp and paper and
5 MMbf to the sawmill/plywood industries and (b) 10 MMbf each to the
pulp and paper and sawmill/plywood industries. Once more, the first
allocation is more representative of Bogalusa and South-wide markets. The
second alternative shifts allocation away from the relatively capital intensive
pulp and paper industry. This alternative should improve the employment
and wage distributive results. The pulp and paper raw material price is
$270/Mbf and the sawmill/plywood raw material price is $290/Mbf.

Results

Table 7.7 reports the results from increasing pulpwood production by
15 MMbf and solidwood production by 5 MMbf. Four forest products
industries show the largest impacts but the finance and service sectors show
important indirect impacts. Wages absorb most of the gain in the service
sector. Property absorbs a larger share in finance. Total employment
increases by 217 persons in all sectors—on a regional base of 130,785. The
forestry sector adds 98 employees, or 45 percent of the small total increase.

Table 7.8 reports the more optimistic impacts from increasing pulpwood
and solidwood production by 10 MMbf each. Final demand and TIO

Table 7.6. Bogalusa region Type I and III output, personal income, and employment multipliers

sector	Type I			Type III		
	output	income	employ-ment	output	income	employ-ment
logging	1.33	1.28	1.70	1.50	1.52	2.28
sawmills	1.74	1.53	1.38	2.12	1.92	1.84
hardwood lumber	1.58	1.41	1.16	2.43	2.15	1.54
plywood	1.63	1.53	1.36	2.05	2.01	1.81
other wood products	1.44	1.36	1.26	1.87	1.80	1.68
pulp & paper	1.46	1.41	1.54	1.66	1.66	2.06
agriculture	1.37	1.67	1.28	1.73	2.87	1.71
mining	1.25	1.66	2.36	1.35	1.95	3.15
construction	1.32	1.27	1.33	1.69	1.61	1.78
manufacturing	1.30	1.56	1.89	1.38	1.80	2.53
food products	1.42	1.42	1.86	1.63	1.83	2.49
textiles	1.23	1.22	1.16	1.82	1.87	1.55
utilities	1.28	1.26	1.32	1.50	1.50	1.77
wholesale & retail	1.22	1.12	1.06	2.09	1.67	1.41
finance	1.24	1.43	1.44	1.42	1.80	1.93
services	1.27	1.18	1.17	1.70	1.50	1.56
government	1.03	1.01	1.01	1.89	1.27	1.35
rest of world	1.00	1.00	1.00	0.99	2.36	1.34
household industry	1.00	1.00	1.00	3.02	1.59	1.34
inventory valuation	1.00	0.00	0.00	1.00	0.00	0.00

Table 7.7. Bogalusa region sectoral impacts from increasing pulpwood output by 12.5 million board feet and sawtimber output by 6.5 million board feet

	million 1982$						# jobs
	final demand	TIO	wages	property income	total income	value added	employ-ment
logging	0.000	1.791	0.355	0.173	0.528	0.534	10.51
sawmills	1.885	2.120	0.593	0.134	0.727	0.757	34.66
hardwood lumber	0.000	0.006	0.002	0.000	0.002	0.002	0.26
plywood	1.885	2.036	0.516	0.126	0.642	0.669	37.78
other wood products	0.013	0.030	0.009	0.002	0.010	0.011	0.62
pulp & paper	2.187	2.220	0.514	0.164	0.679	0.724	17.16
agriculture	0.019	0.277	0.024	0.052	0.076	0.081	4.66
mining	0.002	0.031	0.003	0.013	0.016	0.019	0.07
construction	0.000	0.074	0.024	0.003	0.027	0.028	1.23
manufacturing	0.144	0.733	0.069	0.034	0.103	0.119	1.80
food products	0.100	0.124	0.019	0.010	0.029	0.029	0.82
textiles	0.031	0.037	0.010	0.003	0.012	0.013	1.15
utilities	0.112	0.458	0.120	0.095	0.215	0.232	4.45
wholesale & retail	0.431	0.459	0.211	0.052	0.263	0.331	22.71
finance	0.653	0.882	0.121	0.405	0.526	0.637	6.38
services	0.750	1.085	0.419	0.165	0.584	0.596	23.96
government	0.031	0.066	0.062	0.001	0.062	0.062	3.38
rest of world	0.000	0.000	0.000	0.000	0.000	0.000	0.00
household industry	0.025	0.025	0.025	0.000	0.025	0.025	2.96
inventory valuation	0.000	0.000	0.000	0.000	0.000	0.000	0.00
total	8.267	12.453	3.094	1.431	4.524	4.866	174.58

population change = 519

increase, as expected, over the first Bogalusa scenario. Four forest products industries again show the largest impacts.

Comparing table 7.8 with the base case in table 7.5 shows that total employment in all sectors increases by 307, or 0.2 percent. Forestry sector employment increases by 145, 47 percent of the total employment increase but only a 2.1 percent increase on base case forestry employment. Wage effects are sharpest in four forest products industries and unimportant elsewhere. They range from increases of 0.1 percent increase in other wood products to 1.9 in logging, 7.4 percent in pulp and paper, and 11.3 in plywood. Logging and pulp and paper are among the three highest paid sectors.

In sum, the important distributive impacts repeat the south Georgia experience. They are not great under any scenario. The distributive merit of southern pine research is doubtful for the Bogalusa area.

Table 7.8. Bogalusa region sectoral impacts from increasing both pulpwood and sawtimber
output by 10 million board feet

| sector | million 1982$ | | | | | | # jobs |
	final demand	TIO	wages	property income	total income	value added	employ- ment
logging	0.000	2.536	0.502	0.245	0.747	0.757	14.89
sawmills	2.900	3.206	0.897	0.203	1.100	1.145	52.42
hardwood lumber	0.000	0.009	0.003	0.000	0.003	0.003	0.39
plywood	2.900	3.129	0.793	0.193	0.987	1.028	58.08
other wood products	0.017	0.041	0.012	0.002	0.014	0.014	0.84
pulp & paper	1.621	1.660	0.385	0.123	0.507	0.541	12.83
agriculture	0.026	0.390	0.034	0.074	0.107	0.114	6.55
mining	0.003	0.036	0.003	0.015	0.018	0.022	0.08
construction	0.000	0.091	0.029	0.004	0.033	0.034	1.50
manufacturing	0.196	0.845	0.080	0.039	0.118	0.137	2.07
food products	0.136	0.168	0.025	0.014	0.039	0.040	1.12
textiles	0.042	0.051	0.013	0.004	0.017	0.017	1.56
utilities	0.152	0.557	0.146	0.115	0.261	0.282	5.41
wholesale & retail	0.586	0.623	0.287	0.070	0.357	0.449	30.83
finance	0.887	1.191	0.164	0.546	0.710	0.860	8.62
services	1.020	1.447	0.559	0.220	0.779	0.795	31.97
government	0.042	0.081	0.076	0.001	0.077	0.077	4.15
rest of world	0.000	0.000	0.000	0.000	0.000	0.000	0.00
household industry	0.033	0.033	0.033	0.000	0.033	0.033	4.03
inventory valuation	0.000	0.000	0.000	0.000	0.000	0.000	0.00
total	10.561	16.092	4.040	1.868	5.907	6.346	237.34

population change = 705

FORESTRY IN COASTAL NORTH CAROLINA

Forestry is not a major industry but it has major potential in the twelve counties around the city of New Bern and Beaufort County, North Carolina. These twelve counties contain 2.2 million acres of forestland. Twenty-nine percent of this is industrial, 71 percent is NIPF. The 1984 forest survey reports that 41 percent of industrial forests and 9 percent of NIPF forests are pine plantations. Plantation acreages are increasing for both ownerships, although total forestland is decreasing as demand for forest conversion to agriculture is high. The region supports 76 MMcf of annual softwood removals, 24 percent industrial and 76 percent NIPF.

The forest products industries are also somewhat limited in this region. There is only one integrated pulp and paper mill in the twelve counties—although several in adjacent regions purchase pulpwood from the twelve county region.

The region is most interesting for its forestry potential. Industrial research by Weyerhaeuser and others indicates the potential for large gains in forest productivity from draining the native mixed pine stands and replacing them with intensively managed southern pine plantations. Campbell and Hughes (1980) anticipate gains exceeding 100 percent from drainage alone. Intensive management may raise some sites from site index 40 to site index 80.[5]

Table 7.9 reports economic activity by I-O sector. The combined forest products industries are relatively small, accounting for only 5.3 percent of TIO, 3.4 percent of value added, 2.3 percent of employment, and 3.5 percent of wages. Pulp and paper accounts for 56 percent of the entire sector's TIO but only 26 percent of its employment. Average forestry and forest products wages are moderate for the region—except for the pulp and paper industry whose workers are the region's best paid.

Table 7.10 shows the IMPLAN Type I and Type III multipliers for the twelve county region. They are similar to but smaller than those for south Georgia. The pulp and paper multipliers in particular are smaller in coastal Carolina. All multipliers, nevertheless, fall within the range of Flick's state averages.

Once more, forestry's small share of regional employment but larger regional wage share anticipates that forestry sector investments will not have important distributive impacts. The smaller forestry sector multipliers for coastal Carolina than for south Georgia reinforce this expectation.

IMPLAN Modifications

A 3.0 percent increase in 1984 harvest levels is 12 MMbf. We will examine the alternative allocations of (a) 9 MMbf to pulp and paper and 3 MMbf to the sawmill-plywood industries and (b) 6 MMbf each to the pulp and paper and sawmill-plywood industries. Once more, the first allocation is more representative of coastal Carolina and South-wide markets. The second alternative shifts allocation away from the relatively capital intensive pulp and paper industry. Therefore, it should improve the employment and wage distributive results. The pulp and paper raw material price is $250/Mbf and the sawmill/plywood raw material price is $295/Mbf.

Results

Table 7.11 reports the results from increasing pulpwood production by 9 MMbf and solidwood production by 3 MMbf. Four forest products industries show the largest impacts but the wholesale/retail, service and agriculture sectors show important indirect impacts. Wages absorb most of the gain in the first two. Property absorbs a larger share in agriculture. Total employment increases by 73 persons in all sectors—on a regional base of 214,206. The forestry sector adds 48 employees, or 66 percent of the

Table 7.9. Coastal North Carolina sector and selected share levels

sector	million 1982$							# jobs
	intermediate demand	final demand	TIO	wages	property income	total income	value added	employment
logging	44.80	31.63	76.43	15.13	7.39	22.52	22.81	760.36
sawmills	23.45	32.81	56.26	15.74	3.56	19.30	20.09	693.58
hardwood lumber	2.47	3.24	5.72	1.90	0.17	2.07	2.17	164.21
plywood	16.93	17.05	33.98	9.31	2.23	11.54	11.90	673.05
other wood products	4.61	58.63	63.24	19.72	5.22	24.94	25.51	1,362.48
pulp & paper	9.96	289.86	299.83	49.06	31.67	80.73	83.56	1,264.43
agriculture	367.42	760.19	1,127.61	91.93	328.04	419.97	440.63	18,675.00
mining	1.93	110.35	112.28	23.35	25.68	49.03	52.49	1,287.48
construction	125.15	614.19	739.33	247.67	24.04	271.71	280.05	10,854.12
manufacturing	496.59	1,728.82	2,225.42	593.05	162.41	755.46	785.40	17,883.57
food products	119.20	656.95	776.15	91.40	59.17	150.57	153.58	4,965.31
textiles	94.36	724.61	818.97	187.24	38.39	225.63	232.02	15,615.02
utilities	177.61	189.08	366.68	105.24	86.21	191.45	208.86	5,155.53
wholesale & retail	33.81	429.57	463.38	213.71	52.07	265.78	334.05	30,418.28
finance	183.59	513.46	697.06	74.03	356.62	430.65	526.39	5,397.93
services	84.87	1,065.77	1,150.64	461.91	136.29	598.20	614.67	35,977.76
government	45.58	980.54	1,026.12	946.95	8.36	955.31	955.31	60,722.54
rest of world	0.00	77.03	77.03	(0.14)	77.17	77.03	77.03	(40.14)
household industry	0.00	19.27	19.27	19.27	0.00	19.27	19.27	2,376.21
inventory valuation	0.00	0.00	(17.09)	0.00	(17.09)	(17.09)	(17.09)	0.00
total	1,832.34	8,303.05	10,118.30	3,166.47	1,387.60	4,554.06	4,828.70	214,206.70
wood products %	3.1%	4.8%	4.5%	3.0%	3.1%	3.0%	3.0%	1.9%
agriculture %	20.1%	9.2%	11.1%	2.9%	23.6%	9.2%	9.1%	8.7%
textiles %	5.1%	8.7%	8.1%	5.9%	2.8%	5.0%	4.8%	7.3%
manufacturing %	27.1%	20.8%	22.0%	18.7%	11.7%	16.6%	16.3%	8.3%
construction %	6.8%	7.4%	7.3%	7.8%	1.7%	6.0%	5.8%	5.1%
services %	4.6%	12.8%	11.4%	14.6%	9.8%	13.1%	12.7%	16.8%
finance %	10.0%	6.2%	6.9%	2.3%	25.7%	9.5%	10.9%	2.5%

population = 519,200

Table 7.10. Coastal North Carolina Type I and III output, personal income, and employment multipliers

sector	Type I			Type III		
	output	income	employ-ment	output	income	employ-ment
logging	1.46	1.33	1.71	1.60	1.54	2.07
sawmills	1.77	1.53	1.76	1.95	1.72	2.13
hardwood lumber	1.49	1.36	1.24	1.79	1.62	1.50
plywood	1.55	1.43	1.38	1.78	1.68	1.67
other wood products	1.30	1.26	1.23	1.52	1.46	1.49
pulp & paper	1.13	1.22	1.44	1.18	1.31	1.74
agriculture	1.26	1.56	1.23	1.43	2.17	1.49
mining	1.13	1.20	1.20	1.24	1.36	1.46
construction	1.15	1.13	1.23	1.30	1.26	1.49
manufacturing	1.23	1.25	1.37	1.32	1.35	1.66
food products	1.57	1.59	2.30	1.69	1.89	2.79
textiles	1.37	1.44	1.27	1.58	1.70	1.54
utilities	1.13	1.14	1.17	1.26	1.28	1.42
wholesale	1.10	1.06	1.03	1.66	1.42	1.25
finance	1.14	1.32	1.26	1.22	1.55	1.53
services	1.20	1.13	1.11	1.49	1.34	1.34
government	1.03	1.01	1.01	1.54	1.17	1.23
rest of world	1.00	1.00	1.00	1.00	1.69	1.21
household industry	1.00	1.00	1.00	2.03	1.30	1.21
inventory	1.00	0.00	0.00	1.00	0.00	0.00

small total increase.

Table 7.12 reports the more optimistic impacts from increasing pulpwood and solidwood production by 6 MMbf each. Final demand and TIO increase, as expected, over the first coastal Carolina scenario. Four forest products industries again show the largest impacts.

Comparing table 7.12 with the base case in table 7.9 shows that total employment in all sectors increases by 122, or 0.06 percent. Forestry sector employment increases by 82, 67 percent of the total employment increase but only a 1.7 percent increase on base case forestry employment. Wage effects are sharpest in three forest products industries and unimportant elsewhere. They range from increases of 0.02 percent in other wood products to 1.8 percent in logging, 3.4 percent in sawmills, and 5.6 percent in plywood. These are not among the six lowest paid sectors of the region but logging and sawmills are among the six highest paid sectors.

In sum, the important distributive impacts repeat the south Georgia and Bogalusa experiences. They are not great under any scenario. The distributive merit of southern pine research is doubtful for coastal Carolina.

Table 7.11. Coastal North Carolina sectoral impacts from increasing pulpwood output by 9 million board feet and sawtimber output by 3 million board feet

	million 1982$						# jobs
sector	final demand	TIO	wages	property income	total income	value added	employ- ment
logging	0.000	0.736	0.146	0.071	0.217	0.220	7.32
sawmills	0.885	0.960	0.269	0.061	0.329	0.343	11.83
hardwood lumber	0.000	0.003	0.001	0.000	0.001	0.001	0.08
plywood	0.885	0.954	0.261	0.063	0.324	0.334	18.90
other wood products	0.003	0.005	0.002	0.000	0.002	0.002	0.12
pulp & paper	2.250	2.271	0.372	0.240	0.612	0.633	9.58
agriculture	0.006	0.234	0.019	0.068	0.087	0.092	3.88
mining	0.000	0.000	0.000	0.000	0.000	0.000	0.00
construction	0.000	0.047	0.016	0.002	0.017	0.018	0.69
manufacturing	0.017	0.155	0.041	0.011	0.053	0.055	1.24
food products	0.041	0.057	0.007	0.004	0.011	0.011	0.36
textiles	0.017	0.029	0.007	0.001	0.008	0.008	0.55
utilities	0.035	0.132	0.038	0.031	0.069	0.075	1.86
wholesale & retail	0.082	0.091	0.042	0.010	0.052	0.065	5.95
finance	0.138	0.194	0.021	0.100	0.120	0.147	1.51
Services	0.148	0.189	0.076	0.022	0.098	0.101	5.90
government	0.014	0.034	0.031	0.000	0.032	0.032	2.00
rest of world	0.000	0.000	0.000	0.000	0.000	0.000	0.00
household industry	0.005	0.005	0.005	0.000	0.005	0.005	0.57
inventory valuation	0.000	0.000	0.000	0.000	0.000	0.000	0.00
total	4.525	6.095	1.351	0.685	2.036	2.140	72.32

population change = 175

SUMMARY AND CONCLUSIONS

Distributive impacts can be one justification for public investment. The important distributive impacts are those that improve the condition of disadvantaged populations. In our case, the question is whether southern pine research leads to expansionary productivity increases for low wage industries, thereby increasing wages for poorer families and hiring the unemployed.

We examined this possibility for three important regions of the South: south Georgia and greater Bogalusa, Louisiana, two centers of forestry activity, and coastal North Carolina, a region where research has sharply improved forestry's productivity. We used the US Forest Service's input-output model to examine forward impacts of southern pine research on the logging and forest products industries and on the other sectors of the

Table 7.12. Coastal North Carolina sectoral impacts from increasing both pulpwood and sawtimber output by 6 million board feet

sector	final demand	TIO	wages	property income	total income	value added	# jobs employ-ment
logging	0.000	1.408	0.279	0.136	0.415	0.420	14.01
sawmills	1.770	1.902	0.532	0.120	0.652	0.679	23.44
hardwood lumber	0.000	0.005	0.002	0.000	0.002	0.002	0.15
plywood	1.770	1.907	0.522	0.125	0.647	0.668	37.76
other wood products	0.005	0.009	0.003	0.001	0.004	0.004	0.20
pulp & paper	1.500	1.518	0.248	0.160	0.409	0.423	6.40
agriculture	0.010	0.441	0.036	0.128	0.164	0.172	7.31
mining	0.000	0.000	0.000	0.000	0.000	0.000	0.00
construction	0.000	0.070	0.023	0.002	0.026	0.027	1.03
manufacturing	0.028	0.184	0.049	0.013	0.063	0.065	1.48
food products	0.069	0.094	0.011	0.007	0.018	0.019	0.60
textiles	0.029	0.041	0.009	0.002	0.011	0.012	0.77
utilities	0.059	0.182	0.052	0.043	0.095	0.104	2.56
wholesale & retail	0.137	0.151	0.070	0.017	0.087	0.109	9.93
finance	0.232	0.321	0.034	0.164	0.198	0.242	2.48
services	0.248	0.306	0.123	0.036	0.159	0.163	9.56
government	0.023	0.050	0.046	0.000	0.046	0.046	2.94
rest of world	0.000	0.000	0.000	0.000	0.000	0.000	0.00
household industry	0.008	0.008	0.008	0.000	0.008	0.008	0.95
inventory valuation	0.000	0.000	0.000	0.000	0.000	0.000	0.00
total	5.889	8.596	2.048	0.956	3.004	3.162	121.58

population change = 295

economy. (Backward linkages with labor in timber production are unimportant because of the small producer surpluses generated by southern pine research and because of labor's small factor share in timber growth and management.) Our specific impacts were the impacts of a 3.0 percent research-induced southern pine productivity increase. This is a generous estimate of the impact of a 0.6 percent annual research-induced productivity increase extended over the 3-5 year period of a normal macroeconomic expansion. We checked for sensitivity to alternate allocations of the productivity increase between solidwood and pulpwood.

The results are unimpressive in all three regions and under either solidwood and pulpwood allocation. The most optimistic south Georgia impacts are for a 0.10 percent employment increase for the total economy, and a 1.1 percent increase in forest sector employment. The comparable impacts for Bogalusa are an 0.23 percent increase in total employment, and a 2.1 percent increase in forestry sector employment. The comparable impacts for coastal Carolina are an 0.06 percent increase in total employment, and a 1.7 percent increase in forestry sector employment. These small regional employment impacts are even less important when we consider that they are not immediate, but occur over a 3-5 year period. The comparable wage effects are never important in the six lower paid (of seventeen) sectors in any of the three regional economies.

The lack of substantial employment and wage effects in three regional economies with notable forestry sectors suggest no reason to anticipate more important employment and wage effects elsewhere in the South. In conclusion, the distributive impacts are small in any scenario and the distributive merit of southern pine research is unlikely.

NOTES

1. Type I multipliers measure (direct plus indirect)/direct impacts. Type II measure (direct plus indirect plus induced)/direct impacts. IMPLAN modifies Type II to create a new Type III multiplier with the same objective, but which better accounts for household spending from increased income (Palmer *et al.* 1985.)

2. I-O analysis assumes no price changes within the region of analysis. This assumption is reasonable for local analysis, but not so reasonable for larger regions. The more recent computable general equilibrium (CGE) technique is responsive to prices rather than output availability. It produces models that are more realistic for larger regions, but also more data intensive. Numerous examples of CGE analysis exist for assessing policy and trade impacts in various sectors (Dervis *et al.* 1982, Ballard *et al.* 1985, Hertel and Tsigas 1988). We know of three in forestry (Percy and Constantino 1987, Boyd and Seldon 1991, and Boyd and Newman 1991).

3. These forestry multipliers are generally smaller than the Georgia multipliers of Schaffer *et al.* (1972) or the state multipliers of Flick *et al.* (1980). State multipliers capture additional indirect and induced effects beyond the smaller 57 county region. Therefore, and Schaffer's and Flick's multipliers should be upper bounds for our expectations.

4. One advantage of IMPLAN is its ability to combine counties from different states. This is not possible with standard state-based I-O tables.

5. Drainage also may create important environmental losses. Freshwater drainage into the brackish marshes along the coast damages important shellfish breeding grounds. The land cover change with conversion to intensive silviculture affects native plant and wildlife habitat. The Environmental Protection Agency, the state environmental agency, and various environmental interest groups have brought action to limit drainage and silvicultural manipulation. The final legal outcome is unclear, but it will probably favor environmental interests.

CHAPTER 8

The Net Benefits of Southern Softwood Forestry Research

UNTIL NOW OUR DISCUSSION of southern pine has focused on the gross benefits of research and technical change. These are the appropriate measures for most of the affected timber producers; the sawmill, plymill, and pulpmill consumers; and the surrounding communities, including forest workers and landowners; because these groups receive most of the benefits of technical change. Their wages, invested capital, and land absorb the price and quantity shocks from adjusting to expanding timber growth and management technologies.

Gross benefits, however, are only a partial measure of the productivity effects of technical change on all of society. Net benefits, deducting for the basic research and development costs which make technical change possible, are the complete measure of social gain. For timber resources, the largest identifiable research costs originate with the government. They are not borne by the benefiting populations. Rather, they are distributed as unidentifiable small shares of the general population's total tax burden.

The task of this chapter is to identify reasonable estimates for the costs of southern pine research and development, to compare these with the benefit estimates established in chapter 6 and to draw conclusions regarding the general social merit originating from this research. The research evaluation literature identifies three general inputs, each with associated costs, that contribute to timber research and development: (1) the direct research inputs themselves, as well as (2) extension and (3) some share of education. The direct research inputs have public, university, private (industry) and cooperative university/private components—although only the public share is approximately quantifiable with confidence. Extension and education help the development part of R&D occur faster and more completely. (They also provide a feedback mechanism for expressing to researchers both industry's and the general public's demands for new research.) We will find that even the public share of direct research inputs

173

is large enough to raise doubts about the broad social merit of previous timber research investments—even southern pine research which we anticipate to be among the better examples in forestry and even with generous benefit and conservative cost estimates. The final chapter reflects on the underlying reasons for and the greater implications of this finding.

This chapter progresses through a review of the various cost categories and their historic expenditure levels to summary observations about net benefits and their social implications. Expenditure records are of universally poor quality but they are sufficient for the broad cost estimates necessary to obtain generous measures of the final net social benefits for southern pine research. Benefit measures from chapter 6 plus the summed historical costs from all R&D input categories permit us to make benefit-cost comparisons.

RESEARCH AND DEVELOPMENT COSTS

The same two problems that frustrate identification of the direct effects of technical change on southern pine growth and management also hamper the appraisal of southern pine research, extension, and education expenditures. These problems are: (1) severely limited expenditure records for research and extension and (2) identification of the lag between the initiation of both the research and general education inputs and these inputs' eventual impacts on southern pine production. The overall difficulty of tying individual (or aggregate) research inputs to eventual changes in aggregate forest production relates to both points.

Chapter 6 side-stepped these expenditure and research-productivity lag problems. The search for a fair appraisal of net social gains makes it impossible to avoid them any longer. One approach to the research expenditure problem is to examine first those outlays that are known and to separate out the share of known expenditures that relate to southern pine production. A second step would identify the remaining research expenditures from less reliable sources and, finally, add those shares of extension and education expenditures that support dissemination and application of the new knowledge gained from research effort. This is our approach.

Our estimates in the second step are most subjective. They depend on known levels of research expenditures in some institutions and their suggested general relationships to research expenditure levels in the remaining institutions. The resulting summary expenditure estimates are imprecise. Therefore, they strongly recommend testing all eventual measures of net social gain for a wide range of sensitivity in the less reliable expenditure estimates. Conservative total expenditure estimates will guarantee final results biased to favor southern pine research. This will permit confident judgments only if overinvestment turns out to be our historic observation.

Our solution to the research-productivity lag problem is no more precise. We can anticipate that the lag can be long, as long as a thirty year timber rotation, and even enough longer to include the periods of both the research activity itself and the diffusion of research results. It will be longer, for example, for research on improved seedlings and shorter for fertilization and thinning research. In both cases, research only pays off when the improved seedlings or fertilized trees are harvested as mature timber. We have no basis for a quantified *a priori* assumption regarding the true lag. Our solution is first to sum aggregate research and development expenditures for brief, and then increasing, periods of time and then to compare each sum with our previous estimates for research benefits from the entire period, 1950 to 1980. This will permit conclusions about net social gains for a broad range of short, and then longer, conceivable research-productivity lags.

One outstanding problem remains. We have benefits from 1950 to 1980 and the costs that produced these benefits. Some of our costs, however, may also contribute to benefits accumulating after 1980. The agriculture research literature finds that the benefits due to a particular injection of research expenditures decay over time. Our forest products research evidence records similarly decaying research benefits. Therefore, we might anticipate that southern pine research benefits accumulating after 1980 but due to research expenditures accruing before, say 1950, might be small. Nevertheless, they could be positive. We will assume these (probably) small additional post-1980 benefits are compensated by extending the observed level of annual 1950-1980 research benefits from 0.6 percent to our generous estimate of 1.0 percent, or even 1.5 percent.

Our detailed solution to the research expenditure problem will become clearer as we investigate expenditure history for the six distinct participants in southern pine research and development: (1) the federal government through US Forest Service experiment stations, (2) state forest research organizations, (3) the forest industry, (4) university/industry research cooperatives, (5) forestry extension organizations and (6) educational institutions. Our discussion is entirely in constant 1967 dollars unless otherwise indicated.

Federal Research

Organized federal forestry research in the South, as in most of the country, is of recent origin. Agricultural research may have begun in the 1890s, but it was not until 1920, after the "devastation" of much of the virgin southern softwood timberlands, that fears of impending timber "famine" induced calls for establishing formal forestry research institutions. The "Capper Report" (USDA Forest Service 1920), encouraged Congress to initiate inventory assessments of the nation's forest resources and to maintain and to expand budget appropriations for the forest experiment station system. As a result, organized research began in the South in the mid-1920s with the first field measurements of the southern forest inventory and with the establishment of forest experiment stations in New Orleans and Asheville. The Capper Report also spawned the Clarke-McNary Act of 1924 which established cooperative federal funding for forestry research.

Table 8.1 records expenditures on federal research related to southern softwood growth and management, in 1967 dollars, by the Southern and Southeastern Forest Experiment Stations for the period 1935 to 1980. Occasional breaks appear in these data. There is no record at all prior to 1935 (and we make no attempt either to estimate missing observations from these earlier years or to adjust for their impacts on productivity or technical change).

A full expenditure record since 1935 includes two research categories which may not contribute to southern pine production: expenditures for wood utilization research and expenditures for other miscellaneous research items. The first features wood utilization but probably includes some timber growth and management research. The second surely buries at least a few activities that are relevant to southern pine production. These two categories generally account for 20-25 percent of the annual experiment station budgets throughout the 46 year period of record. Completely excluding them from table 8.1 helps ensure a conservative bias for the eventual full cost accounting. There are no further downward adjustments in table 8.1 for research from other categories that are potentially unrelated to softwood management; for example, hardwood management or non-timber forest production. These data, where they exist, are insufficiently specific. They are also small components of the full federal experiment station research budget.

The main thrust of Forest Service research in the South has had to do with management of southern pine. The expenditure share for research unrelated to southern pine has increased in recent years but it is still a small share of total federal research expenditure. Moreover, the expanding share of southern pine-related research conducted outside the region, yet not reflected in table 8.1, probably exceeds it. In sum, the final column of table 8.1 provides annual estimates of federal expenditures for southern pine research that are reasonable for our purposes. The aggregate bias in these estimates may be small. It is surely conservative.

The table 8.1 history of federal forestry research suggests two general observations. First, the major share of total softwood research expenditures occurred in the recent past. Of total federal softwood research expenditures prior to 1980, 46 percent occurred in the 1970s. Therefore, if public expenditures have contributed to the expansion in timber production over the last three decades, then these recent spending level increases probably anticipate further research-induced productivity expansion in the future. (Of course, this is possible regardless of whether the historical record of

Table 8.1. Southern and Southeastern Forest Experiment Station research expenditures (in 1967$) by functional activity broadly associated with softwood forest productivity, 1935-1980

year	renewable resource evaluation	fire	renewable resource economics	insects and disease	timber management	total softwood research
1935	91,092		47,690		228,392	367,174
1936	261,391		49,520		241,607	552,518
1937	184,270		46,404		317,416	548,090
1938	227,160	3,704	50,988		348,765	630,617
1939	296,482	3,769	51,884		354,899	707,035
1940	296,296	5,556	68,272		331,481	701,605
1941	263,858	4,989	46,896		272,173	587,916
1942	227,898	3,929	41,552		241,159	514,538
1943						450,000ᵃ
1944	97,015	9,328	31,716		171,642	309,701
1945						450,000ᵃ
1946	108,456	21,467	2,978		760,528	893,429
1947	267,190		23,429		750,867	1,041,485
1948	236,901				838,243	1,075,144
1949	247,141				892,915	1,140,056
1950	246,357		21,027		848,985	1,116,369
1951	205,269	30,723	44,457		657,999	938,448
1952	197,227	31,590	27,472		663,995	920,284
1953	230,664		17,162		817,507	1,065,333
1954	216,895	33,079	30,936		691,651	972,561
1955	222,096	38,698	30,182	178,613	753,856	1,223,445
1956	214,994	37,461	29,768	193,045	1,084,577	1,559,846
1957	274,384	82,420	90,139	246,517	1,254,019	1,947,479
1958	360,534	180,021	93,505	267,450	1,350,567	2,252,077
1959	359,735	253,544	93,308	266,886	1,718,267	2,691,741
1960	378,249	268,035	99,357	284,347	1,830,165	2,860,153
1961	381,897	280,977	100,397	371,434	2,011,457	3,146,162
1962	445,530	372,493	163,476	683,550	2,317,803	3,982,850
1963	442,711	376,850	176,323	812,614	2,233,653	4,042,150
1964	469,759	398,229	187,144	1,055,480	2,489,527	4,600,139
1965	523,168	396,915	186,894	1,542,955	2,597,277	5,247,210
1966	518,968	555,611	184,619	1,549,825	2,604,509	5,413,532
1967	579,430	547,200	251,950	1,681,625	2,773,400	5,833,605
1968	626,439	546,049	295,317	1,808,098	2,884,390	6,160,293
1969	623,944	521,784	293,803	1,800,751	2,913,333	6,153,615

Table 8.1 (*continued*)

year	renewable resource evaluation	fire	renewable resource economics	insects and disease	timber management	total softwood research
1970	668,750	535,870	292,391	1,814,583	2,913,406	6,225,000
1971	804,649	547,895	341,053	1,866,053	3,060,877	6,620,526
1972	841,478	581,612	357,683	2,083,375	3,321,662	7,185,810
1973	746,474	906,682	317,372	1,922,717	3,334,001	7,227,246
1974	628,045	650,406	248,282	1,640,162	2,585,572	5,752,467
1975	631,504	636,135	249,571	1,630,532	2,596,512	5,744,254ª
1976	732,514	688,852	250,492	2,170,109	2,753,716	6,595,683
1977	1,046,807	693,872	253,296	2,195,211	2,778,527	6,967,714
1978	1,325,036	670,712	247,014	2,122,504	3,138,892	7,504,157
1979	1,216,469	617,997	227,080	1,952,037	3,140,492	7,154,075
1980	1,094,122	568,080	213,542	1,768,229	2,763,021	6,406,994
1935-60	5,894,589	1,008,314	1,130,013	1,436,858	17,791,378	27,261,151
1935-65	7,974,618	2,833,777	1,852,877	5,902,891	29,071,393	48,535,556
1935-70	11,175,184	5,540,291	3,262,326	14,557,773	43,530,134	78,065,708
1935-75	14,644,298	8,863,020	4,684,918	23,700,612	58,059,056	110,851,904
1935-80	20,242,281	12,102,533	5,967,710	33,908,702	73,003,407	145,224,633

Expenditures converted to 1967$ by dividing the annual values by the All
Commodities Producer Price Index. All data compiled by D. Hair for
the South's Fourth Forest (USDA Forest Service 1988).

ªEstimate.

technical change suggests that southern pine research investments have occurred at a socially justifiable level.)

The second observation, based on the middle five columns of the table, is that the experiment stations have maintained a relatively constant research focus on forest management. Their original emphasis was on two major topics: (1) forest inventory/forest survey and (2) regenerating unproductive cutover pine forestland. These two topics attracted over 90 percent of the total research budget through the mid-1950s. More recent research has added breadth to include more general forest management and the proportion of funds spent on inventory/survey and pine regeneration has decreased somewhat to around 60 percent of the total. Other research interests, particularly insect and disease management (transferred to Forest Service jurisdiction in 1955) and fire, attract increasing attention. The current focus of forestry research is quite broad, including virtually all aspects of southern forest management.[1]

This table of annual Forest Service research expenditures is the only complete data set for any of the six broad participant groups active in southern pine research and development. As such, it includes the basic data upon which the extrapolation of the other participant group expenditures often rests. In the 1930s and 1940s, the total of all regional Forest Service experiment station budgets ranged as high as 90 percent of total timber growth and management research expenditures nationwide. Throughout the 1960s, the experiment station budgets probably still represented more than 50 percent of organized research and development expenditures nationwide. Since then, funds from other sources have increased rapidly until their current share is several times more than the Forest Service share of all softwood research funding. Southern research is now a dominant share of the nationwide total. Therefore, there is no reason to expect that the southern research expenditure pattern has been much different from this national pattern.

State and University Research

For most states, university research is the largest share in this category. Furthermore, the universities conduct a large share of all state supported research. Their budgets may reflect these state funds.

The University of Georgia, in 1912, opened the first university-based forestry school in the South.[2] The other states introduced their own university programs and, since 1930, at least one state-supported university in every state in the region has maintained a professional forestry program. Since the late 1960s, many of the southern states have supported two-year technical forestry programs as well.

The limited available information suggests that these university forestry programs had relatively small research budgets prior to the early 1960s. In 1953, for example, the nationwide contribution to research expenditures from forestry schools was only 28 percent of the Forest Service research budget (Kaufert and Cummings 1955). The southern state universities were relatively undeveloped research institutions at that time. Therefore, their share of all southern research expenditures may have been even smaller.

Since 1953, universities in general and southern universities in particular have increased their forestry research outputs. It is unclear, however, how much of their own funds have been spent on forestry research. In addition to their own research funds, the land grant universities (all southern professional forestry schools except the one at Duke University) have greatly benefitted from the McIntire-Stennis Act of 1962 which provides matching federal support for forestry research at these institutions. This source of funds increased from $0.9 million at southern forestry schools in 1968 to $1.4 million in 1983 (in 1967 dollars)—when it was equivalent to nearly 40 percent of federal funding (or nearly 80 percent, including McIntire-Stennis and the matching university share) for forestry research in the South (Gray 1989). Some McIntire-Stennis funds are used for research topics which have little impact on timber productivity. Nevertheless, university expenditures, including McIntire-Stennis funds, are now a major share of timber growth and management research expenditures in the region.

Sullivan (1977) estimates that the universities received one-sixth of all USDA research funding in the mid-1970s. He further estimates that this was one-third of the total university research expenditure on forestry nationwide at that time. Hodges *et al.* (1988) observe that southern university research expenditures increased to $2.6 million of appropriated funds and $4.5 million of unappropriated funds by the mid-1980s. Hodges *et al.* also observe that southern university research is now almost 50 percent of all university forestry research nationwide.

In sum, state and university research funding has increased relative to Forest Service expenditures, but it has risen to a high level only recently. Table 8.2 summarizes those, often qualitative, expenditure observations available for the various institutional participants in southern forestry research and development. Table 8.3 converts these observations into our final expenditure estimates for the same institutional participants. The first column of each table repeats the Forest Service experiment station estimates from table 8.1. The second column summarizes our state and university expenditure insights. It shows the latter expenditures rising from zero in the early 1950s to a level almost equal to Forest Service experiment station expenditures by 1980.

Forest Industry Research

Industrial forestry research prior to the 1950s was oriented towards finding better ways to manage cutover timberlands. It often featured improved conservation activities and reforestation techniques. Therefore, the forest industry research budget was incorporated in the operating cost of industrial woodlands. It often still is. Woodlands data are usually proprietary, the research component is difficult to separate, and neither may be available from many firms for even one full timber rotation. Nevertheless, many firms with large forest landholdings have set up formal forest (as opposed to product and utilization) research divisions since the 1960s and future evaluations of research expenditures from this source should be more feasible.[3]

Industry expenditures in the early 1950s were only 16 percent of the national forestry research budget (Kaufert and Cummings 1955). Industry research funding in the South increased primarily as a result of the rapid increase in industrial landholdings that occurred throughout the late 1950s and the 1960s. This created an urgent demand for the information necessary to move these lands into higher timber production. Nevertheless, prior to

Table 8.2. Southern softwood research expenditure history, all sources (in thousands of 1967$)

year	(1) USFS research	(2) state and university research	(3) industry research	(4) university industry cooperatives	(5) federal & state extension	(6) general education	(7) total
1935	< 368			0		•	
1940	< 705		very small	0	marginal	•	
1946	< 894	very small		0		•	
early 1950s	~1,000		16% national total	began at NCSU	700	•	
1953	<1,100	<28% USFS total nationwide	rapidly increasing, but still not large	•	715	•	
1968	6,200			200	2,840	•	
mid-late 1970s	7,000	1/6 of USDA - 1/3 of university total	less than 2x USFS total, rapidly declining	20-25% of industry total	6,200	•	all USFS - 35% total research not including extension and general education
1980	6,400	•		•	10,400	•	
mid-1900s	•	2,604 (university), 4,502 (unappropriated)	5,627	900	•	•	24,000 nationwide, not including extension and general education

Table 8.3. Estimated southern softwood research expenditures, all sources (in thousands of 1967$)

year	(1) USFS research	(2) state and university research	(3) industry research	(4) university-industry cooperatives	(5) federal & state extension	(6) general education	(7) total*
1935	367	0	0	0	0	•	380
1940	702	0	0	0	0	•	720
1946	893	0	0	0	0	•	950
early 1950s	1,000	0	180	•	700	•	1,900
1953	1,500	400	300	•	715	•	2,900
1968	6,200	3,200	7,000	200	2,840	•	19,000
mid-late 1970s	7,000	3,800	12,000	•	6,200	•	29,000
1980	6,400	5,500	9,000	•	10,400	•	31,000
mid 1980s	•	7,100	5,627	900	•	•	•

*Totals exclude any estimate for general education. Totals also assume all university/industry cooperative expenditures reappear in either (2) state and university research of (3) industry research accounts.

1970, the level of industrial funding for forest growth and management research (other than for the few university/industry cooperatives, discussed below) was probably quite low.[4] Since the 1970s, however, many firms now fund in-house research units. Total industry spending on all forestry research nationwide peaked in the later 1970s at the annual level of $15-20 million, or two to three times the Forest Service softwood research budget nationwide.[5] The southern proportion may have been dominant because the industry was more actively involved in southern pine research than in research on species more common to other parts of the country. An unknown portion of industrial research, however, was devoted to topics other than softwood forest management, topics like wood utilization.

Industrial research budgets have become tighter in more recent years. The markets for wood products have not been consistently strong and some firms have gone through periods of selling their woodlands. Hodges *et al.* (1988) find that the industry's forestry research budgets declined by 50 percent from the mid-1970s to the mid-1980s. Most of the industry's current research budget is in the South. It was approximately $5.6 million in 1985.

In conclusion, it is difficult to arrive at a fair assessment of the industry input to the total regional expenditure on softwood growth and management research. Expenditure levels that are relatively high compared with Forest Service research expenditures are a recent phenomenon. Column 3 in table 8.2 summarizes this discussion of industrial expenditures. Column 3 in table 8.3 records our dollar estimates.

University/Industry Cooperative Research

Dividing the aggregate gains from research among individual firms may not leave sufficiently large research-induced revenue shares to cover the identical research investments each firm must make in order to keep up with the research breakthroughs of all other firms. (We discussed this justification for either pooling or else public research funding in chapter 1.) Many firms in the forest industry and several universities responded by organizing cooperatives to address specialized forestry research problems. Texas A&M and North Carolina State Universities set up the first cooperatives in the early 1950s for the genetic improvement of southern pines. The activities of these two cooperatives encouraged similar cooperatives to investigate such diverse concerns as nursery management, fertilization, and growth and yield modeling. Total direct funding from the industry to these cooperatives and for other university research increased from $0.2 million in 1968 to $0.9 million in 1983 (Gray 1985). Sullivan (1977) observes that cooperative research expenditures were 20-25 percent of the industry total in the mid-1970s.

These cooperatives have had a profound effect on southern forestry research priorities, and particularly on industrial plantation management. Furthermore, their impacts have spread, as many southern plantation techniques have been adapted outside the South and the cooperative research concept itself now extends into other regions. Seed and tree improvement cooperatives have been particularly well-accepted in other regions of the country. In sum, the productivity effects of the southern cooperatives are widespread. Nevertheless, the southern tree cooperatives' share of the aggregate regional research budget was non-existent before 1950 and it remained small until recent years.

We will err conservatively by assuming that our previous estimates of state/university and industry research expenditures already account for all university/industry cooperative expenditures. Even if this assumption were correct, it is clear that the cooperatives have helped to show industry and university administrators the potential for research to increase timber production. They encouraged an atmosphere conducive to the large expansion of research effort by these other institutions during the past twenty years.

Extension

Forestry extension is responsible for easing the transfer of new or improved management techniques to practicing forest managers. Without it, research would have a delayed impact at best. Boyd and Hyde (1990, ch. 3) and Royer (1985) highlight the significance of forestry extension, particularly in inducing forest regeneration activities among non-industrial private landowners.

Unfortunately, early extension expenditure data are difficult to obtain since the responsible state agencies have generally poor historical budget records. Five of the twelve southern state forest extension agencies responded to our inquiry for budget and related information. Three respondents reported the number of extension workers involved in forest management at ten-year intervals for the last forty years and two respondents shared recent (past five- to seven-year) budgets. These responses indicate great variation across states. For 1980, budgets ranged from $30,000 to $300,000 (current dollars) for softwood forestry. The lower end of this range was common prior to 1970 and generally only marginal expenditures occurred in the 1950s and earlier.

These programs usually receive some form of federal assistance. Federal assistance for all state programs compares with 20 percent of the federal research budget since 1950 (USDA Forest Service 1982, p. 238). The South, because of its good forestry opportunities and relatively large share of non-industrial landowners, receives a substantial proportion, perhaps 33-50 percent of the total. That is, fewer than 25 percent of the states (the twelve southern states) receive 33-50 percent of the national budget. Even in poor years this compares with 14 percent of Forest Service research expenditures in the South and it ranges to 40 percent in high budget years.

The South's Fourth Forest (USDA Forest Service 1988) provides another set of estimates for forestry extension expenditures in the South beginning from the early 1950s. These estimates are consistent with our observations from the previous two paragraphs. They provide our estimates in columns 5 of tables 8.2 and 8.3.

Education

The technical change literature for agriculture shows that general increases in the education level of farmers have substantial impacts on productivity (*e.g.*, Griliches 1963, Welch 1971). Virtually all the forestry literature based on non-industrial private landowner surveys shows a strong positive correlation between the education level of forest landowners and reforestation behavior (*e.g.*, Boyd 1985). Therefore, it is a reasonable argument that forest productivity gains, like those in agriculture, partly originate from improved education. Our accumulation of research and development costs should include the appropriate share of educational expenditures. The problem lies in identifying the share of all southern education that contributes to timber production. Once more, our preference is to err on the conservative side. We will disregard all education costs.

Total Forestry Research and Extension

Column 7 of table 8.3 provides summary estimates of total research and extension expenditures for southern pine growth and management in selected years. The numbers in column 7 are approximately the sums of columns (1) through (3) plus (5); Forest Service, state and university and industry research plus extension, respectively. (Some estimates from tables 8.1 and 8.2 are revised slightly to remove obvious outliers.) These summary estimates are consistent with the summary observations of Sullivan (1977) that USDA forestry research expenditures were 35 percent of the total (excluding extension) in the mid-1970s and Hodges *et al.* (1988) that the 1985 nationwide total was $24 million (in 1967 dollars, excluding extension), most of which was southern.

Our final estimates for each year after 1935 not shown in table 8.3 are the sums of (1) Forest Service expenditures from table 8.1; (2) rough linear extrapolations of (a) state and university expenditures and (b) industry expenditures for years between observations in table 8.3; and (3) extension expenditures reported in *The South's Fourth Forest*. They appear as table 8.4. These final estimates are conservative measures of total R&D expenditures to the extent that they (1) overlook some Southern and Southeastern Forest Experiment Station expenditures on southern pine, including all expenditures prior to 1935, (2) exclude expenditure in other regions of the country which yield spillover benefits for southern pine growth and management, (3) underestimate independent university/industry cooperative expenditures, and (4) overlook the uncertain contribution of general education.

Table 8.5 records these total costs compounded forward to 1980 at annual rates of 0, 4, 7, and 10 percent. Compounding is performed in order to compare total research expenditures with the 1980 present value research benefit estimates from chapter 6.

The annual values in table 8.5 are aggregated into five different year groups, or cohorts, identifying total research expenditures affecting timber production in the period 1950-1980. Surely the earliest research in 1935 still affects production today. Therefore, all expenditure cohorts begin with 1935 expenditures. The most recent research expenditure affecting 1950-1980 production is uncertain. The answer is suggestive of the elusive average research-productivity lag.

For example, if we think the average research-productivity lag is twenty years or longer, then production in 1980, the last year in our benefit calculation, only reflects research expenditures up to 1960; and the first expenditure cohort, 1935-1960, is the appropriate aggregation for comparison. Other cohorts (say 1935-1965) are appropriate for (1) shorter lags (fifteen years) between research expenditures and productivity increases and (2) longer histories of research experience affecting our productivity and benefit measures.[6]

A most conservative measure of total research costs might include only Forest Service expenditures. Surely this is an underestimate of total research expenditures on southern pine growth and management. Therefore, it provides a robust test for the sensitivity of our eventual measures of net social gain to reasonable adjustments in our cost estimates. Table 8.4 also records compounded Forest Service expenditures alone for use in this eventual sensitivity test.

COMPARING TIMBER RESEARCH AND DEVELOPMENT COSTS
WITH THE BENEFITS FROM PRODUCTIVITY SHIFTS

Tables 8.5 and 8.6 provide the bases for drawing net social welfare conclusions about southern pine research productivity. Table 8.5, discussed above, provides estimated total southern pine research and development expenditures. Table 8.6 repeats the previous (chapter 6) estimates of gross benefits from the 1950-1980 southern pine productivity shifts. The chapter 6 estimates show average productivity shifts of 0.6 percent annually. Table 8.6 shows the gross benefit estimates for the reasonable range of annual

Table 8.4. Estimated southern forest research and development expenditures, all sources, 1935-1980 (in 1967$)

year	USFS research	federal & state extension	industry research	university research	total
1935	367,174				367,174
1936	552,518				552,518
1937	548,090				548,090
1938	630,617				630,617
1939	707,035				707,035
1940	701,605				701,605
1941	587,916				587,916
1942	514,538				514,538
1943	450,000				450,000
1944	309,701				309,701
1945	450,000				450,000
1946	893,429				893,429
1947	1,041,485				1,041,485
1948	1,075,144				1,075,144
1949	1,140,056				1,140,056
1950	1,116,369				1,116,369
1951	938,448	463,000	180,000		1,581,448
1952	920,284	504,000	240,000		1,664,284
1953	1,065,333	524,000	300,000	400,000	2,289,333
1954	972,561	534,000	745,000	586,667	2,838,228
1955	1,223,445	592,000	1,190,000	773,334	3,778,779
1956	1,559,846	645,000	1,635,000	960,001	4,799,847
1957	1,947,479	733,000	2,080,000	1,146,668	5,907,147
1958	2,252,077	985,000	2,525,000	1,333,335	7,095,412
1959	2,691,741	1,235,000	2,970,000	1,520,002	8,416,743
1960	2,860,153	1,299,000	3,415,000	1,706,669	9,280,822
1961	3,146,162	1,493,000	3,860,000	1,893,336	10,392,498
1962	3,982,850	1,884,000	4,305,000	2,080,003	12,251,853
1963	4,042,150	1,968,000	4,750,000	2,266,670	13,026,820
1964	4,600,139	2,212,000	5,195,000	2,453,337	14,460,476
1965	5,247,210	2,428,000	5,640,000	2,640,004	15,955,214
1966	5,413,532	2,610,000	6,085,000	2,826,671	16,935,203
1967	5,833,605	2,740,000	6,530,000	3,013,338	18,116,943
1968	6,160,293	2,990,000	7,000,000	3,200,000	19,350,293
1969	6,153,615	3,241,000	7,715,000	3,285,700	20,395,315

Table 8.4. (*continued*)

year	USFS research	federal & state extension	industry research	university research	total
1970	6,225,000	4,245,000	8,430,000	3,371,400	22,271,400
1971	6,620,526	5,166,000	9,145,000	3,457,100	24,388,626
1972	7,185,810	5,816,000	9,860,000	3,542,800	26,404,610
1973	7,227,246	7,456,000	10,575,000	3,628,500	28,886,746
1974	5,752,467	9,239,000	11,290,000	3,714,200	29,995,667
1975	5,744,254	10,571,000	12,000,000	3,800,000	32,115,254
1976	6,595,683	13,090,000	11,400,000	4,140,000	35,225,683
1977	6,967,714	18,130,000	10,800,000	4,480,000	40,377,714
1978	7,504,157	23,170,000	10,200,000	4,820,000	45,694,157
1979	7,154,075	19,861,000	9,600,000	5,160,000	41,775,075
1980	6,406,994	22,370,000	9,000,000	5,500,000	43,276,994
1935-60	27,517,044	7,514,000	15,280,000	8,426,676	58,737,720
1935-65	48,535,556	17,499,000	39,030,000	19,760,026	124,824,582
1935-70	78,321,600	33,325,000	74,790,000	35,457,135	221,893,735
1935-75	110,851,904	71,573,000	127,660,000	53,599,735	363,684,639
1935-80	145,480,526	168,194,000	178,660,000	77,699,735	570,034,261

productivity shifts between 0.5 and 1.0 percent. It also adds the benefit estimates associated with an even more generous 1.5 percent annual productivity shift. This range of values permits us to judge the sensitivity of our final social welfare results against a very broad expanse of productivity shifts and benefit measures.

Table 8.5. **Southern softwood research and development expenditures, 1935-1980 (in 1967$, compounded to 1980)**

cost cohort	U.S. Forest Service expenditures			
	r = 0%	r = 4%	r = 7%	r = 10%
1935-60	27,517,044	88,026,731	216,877,135	547,182,891
1935-65	48,535,556	128,663,353	282,543,069	652,056,193
1935-70	78,321,600	176,301,776	349,632,255	745,789,924
1935-75	110,851,904	219,337,158	402,445,275	810,330,084
1935-80	145,480,526	256,854,632	442,279,982	852,617,356

cost cohort	estimated expenditures from all sources			
	r = 0%	r = 4%	r = 7%	r = 10%
1935-60	58,737,720	164,434,212	363,921,611	826,437,136
1935-65	124,824,582	292,343,234	570,776,751	1,157,054,063
1935-70	221,893,735	447,180,773	788,424,675	1,460,596,066
1935-75	363,684,639	633,070,117	1,015,074,720	1,735,856,986
1935-80	570,034,261	855,844,492	1,251,004,405	1,985,693,151

Values calculated using the formula $V_{1980} = \sum_{t=1935}^{1980} V_t \, (1+r)^{1980-t}$

Table 8.6. Combined 1950-1980 benefits from solidwood and pulpwood stumpage markets (in 1967$, compounded to 1980)

discount rate	productivity shift, annual percentage		
	0.005	0.010	0.015
0%	43,529,902	87,284,148	131,262,731
4%	75,449,424	151,278,810	227,488,147
7%	120,984,431	242,569,250	364,754,435
10%	202,630,770	406,255,683	610,874,701

Values calculated using the formula $V_{1980} = \sum_{t=1950}^{1980} V_t (1+r)^{1980-t}$

Table 8.7 is the summary table. It provides 1980 net present value (NPV) estimates derived from combined cost (table 8.5) and benefit (table 8.6) estimates and a range of discount rates: 0 percent, 4 percent (the rate preferred by the Forest Service), 7 percent (an intermediate rate) and 10 percent (the rate preferred by the federal Office of Management and Budget). These final estimates are all in terms of 1967 dollars compounded forward to 1980. The final estimates could be discounted to an earlier common year with no change in the eventual internal rate of return or

Table 8.7. Net present value for 1950-1980 productivity benefits and 1935-1980 research expenditures from all sources (in millions of 1967$, compounded to 1980)

cost cohort	productivity shift, annual percentage		
	0.005	0.010	0.015
	NPV @ 0%		
1935-60	-15	29	73
1935-65	-81	-38	6
1935-70	-178	-135	-91
1935-75	-320	-276	-232
1935-80	-527	-483	-439
	NPV @ 4%		
1935-60	-89	-13	63
1935-65	-217	-141	-65
1935-70	-372	-296	-220
1935-75	-558	-482	-406
1935-80	-780	-705	-628
	NPV @ 7%		
1935-60	-243	-121	1
1935-65	-450	-328	-206
1935-70	-667	-546	-424
1935-75	-894	-773	-650
1935-80	-1130	-1008	-886
	NPV @ 10%		
1935-60	-624	-420	-216
1935-65	-954	-751	-546
1935-70	-1258	-1054	-850
1935-75	-1533	-1330	-1125
1935-80	-1783	-1579	-1375

benefit-cost determination of the ordinal NPV rankings. Of course, the eventual 1980 NPV is larger than the discounted NPV would be for an earlier year.

The net present value results are negative throughout most of the range of feasible benefit and cost estimates. They become less negative as we move across the table to larger productivity shifts and as we move up the table for smaller discount rates. They become positive only for the briefest cost cohorts—or longest research-productivity lags (these cohorts disregard the final fifteen years or more of research and development costs) and for the extreme combination of lower discount rates and greater rates of technical change. For example, the 1980 NPV is substantial ($73 million) when the estimated productivity shift is a generous 1.5 percent annually, there is no discounting at all, and the research-productivity lag is at least fifteen years. For more likely discount rates and productivity shifts, the NPV is negative regardless of the research-productivity lag.

Previous evaluations of agriculture research generally present their results in terms of an internal rate of return (IRR). The IRR can be positive only when the NPV is positive. Reference to table 8.7 shows that the IRR may be as great as 7 percent—under the most extreme research lag and productivity shift assumptions. Under more likely scenarios, it is always negative.

These results do not speak well for the history of timber growth and management research. Could they be wrong? In particular, could these results be especially sensitive to a change in our understanding of our weakest data, our cost data? Table 8.8 shows the estimated NPVs when we disregard all but Forest Service expenditures for which there are reliable records. This entirely ignores industry, state and university research, education and forestry extension costs. These other institutions conceivably are even more important producers of southern pine research, and more important contributors to southern pine research productivity, than the Forest Service.

Nevertheless, even disregarding all research expenditures except those of the Forest Service does not improve the picture substantially. The NPV is either negative or very small for any positive discount rate and for all productivity shifts within our estimated range of 0.5-1.0 percent annually, together with all research lags less than fifteen years.

The history of southern pine research productivity could be considered from yet another perspective. The analysis in chapter 6 examined research productivity in three periods. The average productivity increase across all three periods was 0.6 percent annually, but it began slowly in the first period (1953-1963), rose sharply to approximately 1.1 percent annually in the second period (1963-1973), then declined to 0.2 percent annually in the

Table 8.8. Net present value for 1950-1980 productivity benefits and estimated US Forest Service softwood research expenditures (in millions of 1967$, compounded to 1980)

cost cohort	productivity shift, annual percentage		
	0.005	0.010	0.015
	NPV @ 0%		
1935-60	16	60	104
1935-65	-5	39	83
1935-70	-35	9	53
1935-75	-67	-24	20
1935-80	-102	-58	14
	NPV @ 4%		
1935-60	-13	63	139
1935-65	-53	23	99
1935-70	-101	-25	51
1935-75	-144	-68	8
1935-80	-181	-106	-29
	NPV @ 7%		
1935-60	-96	26	148
1935-65	-162	-43	82
1935-70	-229	-107	15
1935-75	-281	-160	-38
1935-80	-321	-200	-78
	NPV @ 10%		
1935-60	-345	-141	64
1935-65	-449	-246	-41
1935-70	-543	-340	-135
1935-75	-608	-404	-199
1935-80	-650	-446	-242

third period (1973-1983). The 0.9 percent difference between the second and third periods may be due to an unmeasured external agent that affected all production. It shows up in our residual measure of productivity change, but does not really imply anything about research. The obvious examples are unanticipated levels of fire loss and insect and disease epidemics.

We could construct a new argument that the rate of productivity increase from the first to the second period reflects the initial build-up of research inputs to natural stands. The argument would continue that the rate of productivity increase in the second period is the general case. The decline in the rate of productivity from the second to the third period represents the impacts of unanticipated and unusual events like insect and disease attacks. If the overall rate of research productivity actually continued unabated, then the full continuing research productivity effect might remain at the higher 1.1 percent annually.[7]

This argument encourages us to reconsider the final two columns of table 8.7, with annual productivity increases between 1.0 and 1.5 percent. Nevertheless, even annual productivity shifts of 1.5 percent fail to justify previous southern pine research expenditures except in the extreme cases of short research-productivity lags and rates of return barely 7 percent or less.

In sum, timber growth and management research has not been an outstanding historic investment. This observation is robust. It does not change when we exclude research and development costs from the early years and from all other research institutions other than the Forest Service. It does not change when we generously use only the largest decennial estimate of research-induced productivity gain. Only when we do both, exclude many research costs and incorporate the most generous productivity estimate, do the payoffs to research and development turn unambiguously positive. (See the final column of table 8.8.) This is a truly extreme—and unlikely—case.

SUMMARY AND CONCLUSIONS

This chapter compares the benefits of timber growth and management research measured in chapter 6 with estimates of the research and development costs necessary to provide these benefits. It is difficult to obtain realistic cost estimates and yet more difficult to identify the correct year to year association between benefits with costs.

The chapter 6 estimates of the benefits of technical change, or increased productivity, reflect all the responses to research and its implementation in southern softwood timber production. They reflect public, private, university and cooperative university/industry research, plus those parts of extension and education which are the means of delivering research breakthroughs to the forest landowners and managers who apply them. These chapter 6 benefits reflect a long history of research and development expenditures, a history that precedes our 1950-1980 benefit estimates. This coincides with the substantial anticipated lag between research and the first eventual on-the-ground output increase. Some modern management practices and even more current output may still be the result of research breakthroughs made early in this century. Word of research breakthroughs spreads slowly to some private forest landowners and many technical changes cannot be implemented until the next forest plantation cycle, but then continue to affect production throughout the cycle. For example, the advantages of an improved seedling may accumulate each year throughout its growth period of perhaps thirty or forty years.

Our approach to adjusting for poor information about this mismatch between the timing of research expenditures and their impacts on output begins with accumulating what annual southern pine research cost data is available, primarily from the federal Southern and Southeastern Forest

Experiment Stations. There are also various general statements about the historic expenditure patterns for other research institutions. Some of these general statements also refer to proportional relationships with the known federal expenditures and, thereby, permit us to build approximate expenditure histories for these institutions as well.

This evidence, altogether, suggests an estimate of total southern pine timber research costs for any year. The research-productivity lag remains a problem. Our approach is to accumulate the total costs for increasing five year periods (1935-1960, 1935-1965, etc.) and to compare them with our benefit estimates. Certainly 1935 research breakthroughs still have an impact on production and certainly 1960 research has begun to have an impact—but what about more recent research? We can only hypothesize—and consider the measured results from alternative hypotheses.

In fact, our results are not sensitive even to conservative institutional cost aggregations and intertemporal accumulations. Net present value and internal rate of return estimates are uniformly poor, and usually negative, regardless of the discount rate, the research-productivity lag, or the generosity of the estimate of gross research benefits. Apparently research benefits in southern softwood growth and management have not led to large social gains. These results are disappointing when we compare them with results in agriculture and in the forest products industries. Rates of return on research investments typically range upward from 30 percent annually in agriculture and may exceed 100 percent. We found, earlier in this book, that they greatly exceed 100 percent annually in some forest products industries. Our southern softwood growth and management research results do not begin to approach this range of results and they are particularly disappointing when we consider that the southern pine industry is the most dynamic sector of the timber management industry. If the social gains to timber production research have not been great in this region, then surely they have not been great for other regions of the United States and for other species.

We must consider these (perhaps) surprising results further. Is there any other evidence which supports them? Most timber growth and management research has been of the land-saving or capital (investment time)-saving variety. This is reasonable because land and capital, not labor, are the predominant inputs in timber production. These have not been scarce inputs, however, so long as there have been both old fields converting naturally to timber as well as remaining timber on less accessible sites (*e.g.*, pocosins in North Carolina). If land and capital are not scarce inputs, then no amount of technical change can make production on them less expensive and no amount of research will yield economically productive results.

This has been the history of timber management in the South and Southeast. Indeed, it has been the history of timber management in the United States. It promises to continue to be the pattern in the South for awhile longer as the South is currently going through another period of old field availability and as the decline of the tobacco industry suggests at least one more cycle of eventual old field harvests.

We might predict that research will become a significant contributor to timber production once prices rise to the point where they cover costs of growing the stumpage. Relative stumpage prices have risen through history. Indeed, they are the only basic resource prices which show a long term pattern of relative increases. On the other hand, the best known projections suggest that this pattern may taper off in the early years of the 21st century (Adams and Haynes 1981). The price increase may have already halted in the South (Dutrow 1982). The price increase has been sufficient for prices to now exceed southern pine growth and management costs on the best sites. Therefore, there is potential for research to pay off with production increases and cost savings *on these acres*. If forthcoming long term constant relative prices suggest that the demand price and the cost of growing timber will be in balance on most acres then we can predict that timber growth and management research, while not a success story in the past, will make a larger social contribution in the future.

If so, the pattern of forestry research would not be all that different from that of agriculture research. Ruttan (1980) suggests that it took sixty years before agricultural research really began to make a contribution. Sixty years for forestry would end in approximately 1995. From this perspective, it is possible to view current and past timber management research as the initial capital investment necessary to permit (socially more desirable) future timber production breakthroughs.

NOTES

1. H.R. Josephson (1989) provides an exhaustive discussion of the successes of these research programs and the individuals who led the research.
2. The first forestry school in the US, the Biltmore Forestry School in North Carolina, opened in 1898.
3. Several companies responded to a survey of research expenditures conducted for this analysis. Some provided detailed lists of their previous few years' research budgets. This suggests that future researchers may be able to build more accurate estimates of industrial research.
4. R. Weir, Director, North Carolina State University-Industry Tree Improvement Cooperative, personal communication, 1985.
5. B. Malac, Technical Director, Woodlands Division, Union Camp Corpora-

tion, personal communication, 1986.

6. In this analytical structure, *shorter* research-productivity lags are suggestive of improved research effectiveness. They also mean, however, longer histories of costs which actually contribute to our measured aggregate 1950-1980 productivity increases from chapter 6, therefore greater costs and a final conclusion, in this chapter, of *less* economically efficient research.

7. Another explanation for the declining rate of research productivity might be changing weather, perhaps even acid deposition.

CHAPTER 9

Summary, Conclusions, and Policy Implications

SOLOW (1957) AND DENISON (1974) observe that research—through its product, technical change—has explained between 70 and 80 percent of the increase in US non-farm production in this century. Knutson and Tweeten (1979) observe that the marginal return on agricultural research and extension expenditures in the US has been greater than 35 percent annually since 1939. These impressive observations confirm an historic record of research activities strongly improving aggregate US economic growth. The Knutson-Tweeten observation also implies that there would have been general social gain from even larger investments in agricultural research.

Our question in this book is whether this broad and general research experience has also been true for the forestry sector of the US economy. The purpose of this chapter is to summarize our inquiry into *public* research productivity in the forestry sector. Our intention has been to apply appropriate economic techniques (1) to assess various forestry research programs, with the more important overall objective of (2) providing policy insights into (a) the distinguishing characteristics of those research programs with the strongest prospects for substantial future net social benefits and (b) the technical, institutional, and market bottlenecks that deter even greater research success.

Our general approach is through a select series of empirical cases. These cases include four examples from the forest products industries and one from the timber growth and management industry. The forest products examples are the softwood plywood industry, an industry which is generally thought to be an outstanding example of research success; the sawmill and

The original draft of this chapter was prepared while we were guests of I. Vertinsky and the Forest Economics and Policy Analysis Research Unit at the University of British Columbia in January 1989. FEPA provided financial support. M. Fullerton, T. Heaps, P. Pearse and especially A. Scott raised questions or contributed insights which improved both our analysis and our exposition. R. Sedjo, Resources for the Future; D. McKenney and J. Davis, Australian Centre for International Agricultural Research; and D. Brooks, US Forest Service reviewed and improved subsequent drafts.

woodpulp industries, two industries which largely define the forest products sector of the US economy; and the wood preservatives industry, an industry which provides the special example of product-altering (or quality-oriented) rather than process-altering research. The timber growth and management example is the southern pine industry. The southern pine industry has supported the most successful timber management and biological improvement research in North America. It is the basis of the most advanced timber production region of the US.

A glance ahead to our results shows great gains from forest products research, gains sometimes exceeding the best observed in agriculture. The success of southern pine growth and management research, however, is altogether more doubtful. Because southern pine research is thought to be a good example, this doubt encourages even greater skepticism regarding the general performance of all less promising examples of timber management research in other regions of the US. It also begs two questions: (1) what explains the difference in performance between forest products research and timber management research and (2) what would it take to alter the less satisfactory timber management research history? The major part of this chapter reviews our empirical cases and reflects on the causes of productivity differences between forest products and timber management research.

Our general inquiry was important for the US when we began in 1983 because American forestry research had not previously undergone a technical economic appraisal and because forestry was embarking on a period of unusually tight research budgets. Tight budgets beg careful justification of all expenditures. Furthermore, many developed countries shared this experience of tight budgets. Therefore, many developed country forestry institutions may also gain insight from our US observations.

Since 1983 there has been an even greater need in forestry for a broad understanding of both the technical methods of research evaluation and the most general lessons gained from reflections on previous American forestry research experiences and their implications for the future. This greater need resides in the rapidly expanding forestry research programs in developing countries—and in expectations for forestry research to contribute solutions to the newer global issues of biological diversity and climate change.

Forestry has become an important issue for the global environment and an exciting topic in economic development. An immense amount of money is being spent on forestry, including forestry research, in developing countries today.[1] Much of this money originates with international lending agencies which generally require repayment regardless of the eventual success of local investments. Thus, while forestry research programs intend to produce great gains for the rural poor in many developing countries, they may promise increased and extended national indebtedness instead if these programs are chosen with less than the greatest care and best economic foresight.

An important intention of this final chapter is to bring these public policy concerns together. The technical assessments of research in the five specific forest industries should permit reflection on the characteristics of other, greater and lesser, American successes of the recent past and they should provide some intuition with respect to the future of forestry research in the US. We intend that they should also provide initial justifiable extrapolations for the more general forestry research situations occurring around the world—regardless of the level of local economic development.

There are two basic parts to this chapter. The first reviews and draws the immediate conclusions arising from previous chapters. This review begins with background comments on research evaluation in general and on prevailing opinion regarding the public welfare contributions of research investments. It continues with the summary statements and comparisons of the technical and empirical research assessments for our five industries, as well as a reflection on the economic intuition underlying our results. The second part of the chapter generalizes from these experiences and draws our own implications for American and for world public policy.

BACKGROUND ON RESEARCH PRODUCTIVITY

Two very different items are necessary background preparation for our empirical analyses. The first is a basic understanding of the technical method for measuring net research benefits. The second is a general perspective on the history of research success and technical change in the US economy. This section of the chapter reviews each in turn.

The causal chain from research leads through new knowledge and the resulting technical change and, finally, to increased production. There are no convenient summary measures for new knowledge. Therefore, the common estimate of research output is the increase in productivity due to technical change. There are two kinds of technical change; technical change associated with new *processes* that decrease the costs of producing familiar products, and technical change that improves *product* quality or creates new products including technical change that reduces the level of environmental residuals (a non-market product) associated with a known (market) product. Common estimates of process or cost-reducing change reflect downward shifts in the supply function or outward shifts in the production function for a known product. There are no easy means for estimating product-altering technical change. Therefore, our measures of process research benefits and technical change suffer as underestimates of the full impacts of research to the extent that product quality—and environmental residual—research are important.

Modern econometric estimates of the benefits of research begin with research expenditures as one productive input. This approach permits the direct estimation of (embodied) technical change as the expansion in output explained by research expenditures.[2] It requires sharp attention to research cost accounting and to the lag between research expenditures and increased productivity. For such analysis, research expenditures must include all the costs of making the research productive: the costs of research itself, of communicating the results of successful research to those who may use it, and of modifying existing technologies in order to incorporate it operationally. Therefore, research expenditures include all research and development costs, public and private, and also includes extension expenditures as the cost of communicating research results in industries like agriculture and timber management.

The research term, in statistical estimations, also reflects the research-productivity lag. This lag may be short for many industrial processes but it may also be as long as a timber rotation, or longer, where information transfer is slow and production runs are long. We can anticipate that (1) estimating this lag is a critical part of the research evaluation process and that (2) long lags, together with a positive opportunity cost of capital, are disincentives for some timber management research investments.

Solow (1957) and Denison (1974) and Knutson and Tweeten (1979) used variations on this improved approach to explore process change in the non-agriculture and agriculture sectors, respectively, of the US economy. They observe the impressive results reported at the outset of the chapter. Various others observe returns in the range of 35-110 percent for select research programs. [See Ruttan (1980) and Brumm and Hemphill (1976) for surveys.] Since annual research gains of 35 percent imply a doubling of social gains approximately every two years, these previous empirical analyses suggest that the dynamic gains from research and technical change can rapidly offset the static negative effects of most administered public policy regulations.[3] Furthermore, marginal rates of 35 percent are much greater than the (not strictly comparable) marginal returns on most private investments. This last point suggests the possibility that larger historical research investments could have produced even greater net social gains. Altogether, this experience raises two questions:

- If greater returns were available from increased research investments, then why were research investments not larger?
- Were returns to forestry research as great as returns to research investments elsewhere in the economy?

Research investments were not greater because of the public good nature of some research. The private sector in general and producers in particular do not receive the full net social gains from research. There are three cases where public investment is necessary before research expenditures can reasonably expand to the socially optimal level:

- where research requires large initial investments, where there is a long investment horizon before productivity results are forthcoming, and where these results are highly uncertain;
- where industry demand is highly inelastic or supply is elastic; and research benefits are rapidly and largely transferred to competitive higher level producers and to final consumers;
- even where demand may be more elastic or supply more inelastic, many firms may share the positive aggregate production gains due to research. Therefore, the individual firm's gains are insufficient to cover the full research costs, yet each firm must invest these full costs independently in order to obtain any part of the industry's total gain.[4]

The first class of cases is probably more familiar to the space and defense industries than to forestry. The second class describes the sawmill and woodpulp industries. The third class again describes the sawmill industry, and also the timber management industry, and it may describe the softwood plywood industry. Our empirical analyses examined the justifications for a public research presence in all four of these industries.

FOREST PRODUCTS RESEARCH

Our empirical analyses fall into two categories: forest products research and timber growth and management research. Our prototype evaluation for the forest products industries is the softwood plywood (SWPW) industry, an industry generally thought to be a research success story. Data for the SWPW industry, however, create a requirement for some modification of the accepted general approach for research evaluations. Our modified approach begins with the dual of the production function and with research as an explicit factor of production. We transform this production function to its counterpart supply function and simultaneously estimate supply with demand. This is a data intensive approach relative to most agriculture research evaluations and all previous forestry research evaluations, but it has the advantage of improved accuracy. It also permits eventual calculations of both marginal and average returns to research.

Softwood Plywood Results

Table 9.1 repeats the non-linear two-stage least squares (NL2SLS) estimates for the SWPW demand and supply coefficients for the period 1950-1980. The general statistical tests for these equations are all satisfactory. All individual coefficients have the anticipated signs and all are significant at the 10 percent level or better. The high R^2s indicate that these equations explain most of the variation in the dependent variable. The low Durbin's h statistic shows that serial correlation is not an important problem for the two lagged endogenous production/supply terms. These statistical results provide confidence as we proceed in the SWPW research analysis.

A separate statistical test rejects the hypothesis that the industry experienced other (disembodied) technical changes that are unexplained by either the public or the private R&D terms.[5] This is not surprising. Public R&D, largely at the Forest Products Laboratory (FPL) of the US Forest Service, was extensive in the 1950-1980 period. The FPL participated in virtually all recognized SWPW technical change: new gluing techniques, experiments with little-used log species, and all new capital equipment experiments originating in the US. Private firms, in the SWPW case,

Table 9.1. NL2SLS estimates of SWPW demand and supply coefficients

demand		production/supply	
intercept	8.8088** (3.8696)	labor	.2057** (.1125)
own price	-2.7034* (.4901)	capital	.1263** (.0604)
construction wage	-.6076*** (.4205)	(private R&D)$_{t-2}$	.0660* (.0218)
cost of capital	.1263** (.0597)	(public R&D)$_{t-2}$	.0247** (.0131)
substitute (lumber) price	3.0603* (.4265)	lag	.8694* (.0378)
final product price	2.3583*** (1.5586)		
R^2	.8869		.9839
Durbin-Watson	1.46		- - -
Durbin's h	- - -		.0270
degrees of freedom	22		23

Numbers in parentheses are standard errors.
   * Significant at the 1 percent level
  ** Significant at the 5 percent level
 *** Significant at the 10 percent level
All coefficients implicitly refer to year t except where otherwise indicated.

concentrated their R&D efforts on recognizing, receiving, modifying, and developing public research for the specific needs of individual plywood firms or plants.

The key results in table 9.1, are the last three terms in the production/supply equation. Our public policy focus requires the segregation of public from private research costs, with terms in the production/supply equation for both. The subscripts following the public and private R&D terms refer to the number of years following a research expenditure but preceding its initial impact on output.[6]

The elasticities deriving from the coefficient estimates in table 9.1 are:

supply price	0.497	initial public R&D output	0.0247
demand price	−2.703	long-run public R&D output	0.19
labor output	0.2057	private R&D output	<0.48
capital output	0.1263		

These elasticities are not inconsistent with related previous literature (McKillop *et al.* 1980, Rockel and Buongiorno 1982). All but the private R&D elasticity are significant at the 10 percent level or better. The private output elasticity is a residual. (Its upper bound is one minus the other output elasticities.) Its estimate declines if we anticipate larger scale effects in production.

R&D has an initial effect in the period immediately following its implementation and an accumulating longer-term effect as the research breakthrough deteriorates over time. The initial and the long-run research elasticities reflect this distribution. The *lag* term in table 9.1 explains the difference between the initial and long-run public R&D elasticities. A lag coefficient closer to one indicates that the continuing effects of a single R&D injection endure over a longer period of time.

The market equilibrium price and quantity and the supply price elasticity explain industry production prior to the public research investment. The public R&D elasticity, the two-year research-productivity lag, the subsequent Koyck distribution of research impacts, and the lag coefficient explaining the decline in these impacts over time, altogether determine the position of the new industry supply curve at any time after the R&D injection. The area between the original supply curve and the new supply curve, and to the left of market equilibrium, measures the gross benefit from public R&D.

The first column of table 9.2 shows various final measures of these gross social benefits of public SWPW research. The net present value (NPV) estimates show that primary producers and their higher level industrial processors (consumers) share almost equally in these research benefits.

The large producer gains cause us to inquire about the public presence in this research effort. Was it justified or was there sufficient incentive for the industry to make the necessary research investments itself? The competitive nature of the industry, and the non-exclusive nature of many SWPW research products, suggest that the individual firm's only incentive is the payoff the firm captures before its competitors adopt the same or similar new technologies. This is probably the payoff of the first two years. The annual flows of research benefits and costs, however, indicate that the average firm conducting independent research would not have recovered its research costs in any of the first eighteen years of our analysis—or until 1968.[7] Thus, without a public presence, there would probably have been little research and little gain, for either producers or consumers. We conclude that the public presence in SWPW research in this period was socially advantageous.

The value of the marginal product (*VMP*) and the rate of return estimates further display the efficiency of public SWPW research. The *VMP* measures the gain in output due to the last public research dollar spent. The short-run *VMP* ($12.49) is comparable to those found in agriculture and it shows that additional SWPW research investments would have returned more than their costs. The long-run *VMP* ($58.17) and the marginal internal rate of return (*MIRR*, 299 percent) are greater than usually found in agriculture—perhaps because the initial SWPW research

Table 9.2. Returns to public research investments (1950-1980) in selected forest products industries

	softwood plywood	sawmills	woodpulp	wood preservatives
NPV (consumers' surplus)[a]	$1.32	38.97	$.018	$.062
(producers's surplus)[a]	$1.50	-17.56	-$.022	-$.068
(net economic benefit)[a]	$2.84	25.96	$.004	$.252
IRR on NPV (consumers' surplus)	326%	34%	33%	multiple solutions
(net economic benefit)	499%	28%	15%	293%
VMP (short run)[b]	$12.49	----	----	----
VMP (long run)[c]	$58.17	$2.62	$.44	$.34
MIRR	299%	16%	<0	<0

[a]In billions of 1967$ discounted at 7 percent. Consumers' and producers' surpluses are <u>each</u> net of total research costs. Therefore, net economic benefit - net consumers' plus net producers' surpluses <u>plus</u> total research costs.
[b]In 1967$
[c]In 1967$ discounted at 7 percent

investments were small; or because the SWPW research-productivity lags were so short, therefore, the final research benefits are discounted less. The short-run and long-run *VMP*s and the *MIRR* all indicate that additional net social gains would have been available from even greater investments in SWPW research.

Finally, it is always wise to examine exceptional results for sensitivity to critical features of the analysis. Doubling the estimated private R&D costs and increasing the social discount rate to 10 percent still creates research results that are more impressive than most found in the agriculture literature.

One might question whether our extreme results might also be either a contrivance of our approach or an accident of the industry's geographic shift from the Northwest toward the South in the late 1960s. Southern wages were lower and it only recently became possible to process smaller southern logs for plywood. It turns out that neither of these differences alter our general SWPW conclusions. First, altering our approach to more closely mimic previous research evaluations—which rely on exogenous estimates of the price elasticities—only increases the calculated efficiency of SWPW research. Second, our most complete analytical model permits annual adjustments for exogenous shifts in non-research input costs like wages. Therefore, there is a control on the effect of declining wages, and declining wages do not alter our observations of research efficiency. In sum, there is good reason for confidence in both our SWPW demand-supply estimates and our public research productivity conclusions.

Research in Other Forest Product Industries

We repeated this inquiry into public research investments for three additional industries: sawmills, woodpulp, and wood preservatives. We will summarize our results for those industries and reflect on the differences in research results across all four industries, including SWPW. The important questions have to do with the differences between industries and with explaining what these differences suggest that might be instructive for public research managers and for the allocation of future forest product research funds.

The sawmill analysis furnishes most satisfactory statistical results—on the supply/production and the demand equations and on all coefficients in both equations. The supply and demand price elasticities, 0.34 and −0.60 respectively, are significant at the 1 and 10 percent levels respectively. Both are consistent with previous literature (Robinson 1974, Adams and Haynes 1980, Lewandrowski 1989). The long-run public research output elasticity of 0.93 is significant at the 5 percent level. The research-implementation lag is a notable five years in the sawmill case. This longer lag is probably due to the large number, small size and diffuse nature of firms in the US sawmill industry (17,000 in 1950, 6,800 in 1980). It is difficult to communicate research results quickly with so many small and diffuse clients.

Average returns to public research in this industry fall in the range of 17-57 percent annually for consumers and 13-49 percent for society as a whole—depending on our estimates for the private implementation costs of new technologies. Producers are net losers. This means that, for producers, decreased production costs are largely passed on as lower consumer prices, causing producer revenue losses that cannot be offset by gains from increased outputs. It also means that sawmill managers have little incentive to invest in research and that the social gains from sawmill research would not be forthcoming if we relied on private industrial research. Marginal returns fall between 5 and 32 percent annually. This probably equals or exceeds the social opportunity cost of capital and it means that socially justifiable sawmill research investments were at least as great as, and perhaps greater than, the historic 1950-1980 investment level.[8]

The woodpulp analysis provides satisfying supply results but a problem for demand. Statistical tests on both the woodpulp supply/production equation and on its individual terms are satisfactory. The supply price elasticity of 1.09 is significant at the 1 percent level and the long-run public research output elasticity of 0.09 is significant at the 10 percent level. Our econometric estimation of woodpulp demand is unconvincing, however, and a perfectly inelastic form would better explain the woodpulp-consuming paper industry. Inelastic demand is consistent with both prior woodpulp literature (Guthrie 1972, Gilless and Buongiorno 1987) and more casual observations of the high fixed cost nature of the paper industry. We employ our supply elasticities and an assumption of inelastic demand in our calculations of woodpulp research benefits.

Average returns to public woodpulp research are 33 percent annually for consumers and 15 percent for society as a whole. Producer gains are negative, as they must be in an industry with perfectly inelastic demand. Marginal returns are negative, indicating overinvestment in woodpulp research.

It turns out, however, that some woodpulp research dealt with environmental residuals. This research is of the product-altering, rather than the cost-reducing variety. Therefore, our measures of average and marginal returns do not reflect its benefits. This means that our average and marginal returns are lower bound estimates for the true measures.

Higher government authority mandated some 1970s research on energy-saving and pollution-reducing pulping processes. Pursuit of this research was contrary to the professional judgment of FPL scientists regarding its technical and economic feasibility and this research probably would not have been conducted in the absence of political intervention. The average and marginal returns to all remaining woodpulp research for the thirty-year period would have been higher. This is a second reason why our woodpulp results are lower bound estimates for the performance of FPL scientists and managers.

The wood preservatives analysis is also problematic. Our demand specification is statistically significant but the supply/production equation is less reliable. The demand price elasticity of -1.62 is significant at the 1 percent level. The supply price elasticity of 0.047 is insignificant. (There is no comparable econometric wood preservatives literature.) The R&D coefficients have the correct signs but only the private R&D coefficient is significant. The insignificant public coefficient may be due to the large investment in FPL research designed to reduce the petroleum-based environmental residuals associated with wood preservatives. Public research cost data include this investment but our output measure does not reflect their research products. The resulting partial cost-output mismatch may explain the statistical insignificance.

Nevertheless, we proceed on the confidence of the correct sign on the public coefficient (as is common in the research evaluation literature where statistical insignificance has been the norm). The long-run public research output elasticity is 0.0274. The average social return on public wood preservatives research may be as great as 293 percent annually. (Rates of return to consumers and producers are indeterminate because their time streams change signs several times.) Marginal social returns are negative, indicating either overinvestment or significant benefits in unmeasured product-altering wood preservatives research. Removing the costs of product-quality and environmental research would raise the marginal returns, perhaps to the positive range where they would also remove the question of overinvestment.

Comparative Results

It should be interesting to consider the similarities and differences in research production across the four forest products industries. Table 9.2 summarizes. The first summary observation is that public research investment was socially justifiable at some level in all four industries. Yet:

- The benefits of research would be spread too thinly among eighty establishments to sustain much *private* research investment in softwood plywood. Research benefits would be spread even more thinly among 6,800 or more sawmill firms. Indeed, the sawmill industry conducts virtually no private research of its own.

- Inelastic demands in sawmills and woodpulp indicate that these industries pass most of their research gains to consumers. One reason why concentrated industries, like the woodpulp-consuming paper industry, might vertically integrate is to absorb those consumer benefits.
- The producers of wood preservatives and woodpulp cannot capture the benefits of those research improvements which decrease non-market environmental residuals occurring in the production process.

In sum, all four industries can justify public research at some level on the basis of the resulting consumer and net social benefits. The minimal (often even negative) and intermittent producer benefits, the inability to establish lasting proprietary rights to these benefits, and the unlikelihood of sharing research costs across firms make large private producer investments in research unlikely in these industries. Moreover, these conclusions are insensitive to reasonable adjustments in either private research implementation costs or the social discount rate.

Measures of marginal returns indicate historic underinvestment in softwood plywood and sawmills, and potential overinvestment in woodpulp and wood preservatives. Smaller investments are more difficult to measure precisely. This may explain overinvestment in woodpulp. Conjectured non-measured benefits from product-quality improving research in woodpulp and wood preservatives would boost both average and marginal returns in these industries. This boost may be sufficient to place the marginal returns in the range of social efficiency.

The generalized determinants of research performance are unclear. The results from these four industries suggest that neither firm size nor market power are correlated with public forest products research success. Average returns are greatest in softwood plywood and wood preservatives, which are the middle industries with respect to firm size and market power. The small size of the research programs in softwood plywood and woodpulp, the long research-implementation lag in the diffuse sawmill industry, the absence of unaccounted product-quality benefits in our benefit measures for wood preservatives and perhaps woodpulp, and ill-conceived political encouragement of some woodpulp research all appear to be more important explanatory factors of research performance than either firm size or market power.

In conclusion, public research performance in these four industries for the period 1950-1980 ranged from adequate to superior. It brought social gains that were often greater than those usually anticipated for marginal private investments, comparable to (if more variable than) the 35-110 percent average returns in Ruttan's (1980) survey of public agricultural research investments, and substantially in excess of either (negative returns for) public investments for non-industrial private forest incentives (Boyd and Hyde 1989, ch. 3) or (small positive returns for) public forest timber management (Boyd and Hyde 1989, ch. 8; Repetto 1988). These comparative observations are implicit compliments for the forest products research managers of the 1950-1980 period of our analyses and they should encourage the research managers of today to sustain their best judgments.

TIMBER GROWTH AND MANAGEMENT RESEARCH

Evaluating timber growth and management research poses new problems unfamiliar in either agriculture or forest products research. First, forest landowners, unlike SWPW millowners for example, are not a homogeneous group, all with similar production functions. Rather, the classes of public, industrial, and non-industrial private (NIPF) landowners may each have different management objectives and only industrial landowners may have strict timber production objectives. This restricts our confidence in the profit maximization assumption underlying any specification of an aggregate production function.

Furthermore, there are two new difficulties for direct estimation of stumpage research shifters: (1) The long and uncertain lags between initial research and eventual productivity surely obscure statistical reliability. The lag between *research* and *implementation* may be many years where the process of transmitting information to NIPF landowners is slow. The lag can be even longer where *existing* timberstands are too old to respond substantially to new silvicultural treatment but too young to harvest. In this case, implementation cannot begin until the next timber rotation. The lag between research *implementation* and increased *productivity* can be even greater yet. Eventual productivity increases will be unavailable until some distant future harvest. Moreover, even if landowners quickly adjust for rising expectations, both risk aversion and the uncertain increases in distant future harvests surely decrease those landowners' immediate harvest adjustments and delay their full adjustments for research-induced expected harvest increments. Finally, (2) less frequent data (forest inventories are generally collected in ten year cycles) and data that cannot segregate growing stock from final output put timber management research evaluation at additional disadvantages relative to either agriculture or forest products research evaluation.

This all causes us to reflect further about the dynamics of the research impact on timber production on the stump. We propose, first, to observe the production impact of research and technical change on periodic standing forest inventories and, second, to use forest inventory as an independent variable in the stumpage supply function. Research-induced shifts in inventories alter supply and, together with a known demand, provide evidence of research benefits in financial terms.

In the southern pine case, our production relationship explains standing inventory as a function of various proxies for capital and labor inputs and a time shifter. The time shifter is a proxy for disembodied technical change—measuring those otherwise unaccounted increases in forest inventory that occur regularly with time. The time shifter includes research-induced inventory shifts and it may include some regular increases in productivity due to unspecified causes other than research.

Nevertheless, the statistical estimations for the production relationship are satisfying. The signs on all coefficients in the basic equation support intuition and most are statistically significant. The equation test statistics are also satisfying. Most important for our analysis, the technical change shifters are always significant at the 5 percent level or better, for the basic equation and for various specialized variations on it.

In sum, we have confidence in the ability of the time coefficients to explain shifts in production. They explain approximately 7, 11, and 2 percent shifts in successive decennial inventory periods from 1953 to 1983. The total increase is 20 percent for the full period, or about 0.6 percent annually. This may be a generous estimate of research-induced productivity shifts because it includes both (1) non-research-induced shifts and (2) movements along, as well as shifts in, the production function. It also may be conservative, however, because our approach controls for shifts in forest management type that may be partially research induced. Sensitivity tests for variation in the estimated production shift protect against these biases affecting our final general conclusions.

Our conversations with university forestry researchers and forest industry research administrators suggest that they look for volume increases of 20 percent over a southern pine timber rotation of 25-40 years. Their expectations are consistent with our observations. These are *volume* measures, however. Measures of research *value* must await our market analysis.

Stumpage supply is a second and different function of the various productive inputs, including inventory. The share of standing inventory defined by the coefficient on the time shifter in the production function becomes a supply shifter. Supply is more complex, however, because southern pine production is a joint input for two supply functions, solidwood and pulpwood. Therefore, southern pine research has an impact on two products and we must simultaneously estimate supply and demand for both products, each with a standing forest inventory term.

Table 9.3 repeats the results of our three-stage least-squares (3SLS) demand and supply estimates. (There is no previous econometric evidence of the joint solidwood-pulpwood market for southern pine stumpage.) The general statistical tests for the four equations are satisfactory. All coefficients except one (capital in the solidwood demand equation) have the expected signs.[9] All except two are statistically significant at the 10 percent level. The F and R^2 statistics indicate good equation fits.

The inventory coefficients in table 9.3 are the important values for our analysis. They imply solidwood and pulpwood supply elasticities of 0.391 and 1.201, respectively. These elasticities anticipate that the gross benefits arising from research and technical change are not great because the solidwood inventory elasticity is low, therefore inventory increases have little impact on the solidwood market, and because a much smaller pulpwood price offsets the larger pulpwood inventory elasticity.

Our measure of gross research benefits follows a derivation by Hertford and Schmitz (1977). It relies on our estimate of research-induced inventory shifts and the various demand and supply elasticities. Table 9.4 shows the *undiscounted gross* research benefits accruing to consumers, producers, and all society in both the pulpwood and solidwood markets for the range of research-induced inventory shifts between 0.5 and 1.0 percent annually. Total social benefit estimates are sensitive to this inventory shift but they are insensitive to large variations in the demand and supply elasticities.

Total benefits (for pulpwood and solidwood and for a 1.0 percent annual research-induced productivity change) are small: only $40-80 million, or approximately $200,000 per state per year (1980$), for twelve southern states and the thirty-year period from 1950 to 1980. Consumers (loggers, millowners, and final product consumers) gain more than stumpage producers. This distributive result is similar to the experience of agriculture, sawmills, and woodpulp where research benefits quickly pass to higher level producers and final consumers in the form of lower prices. The emphasis on consumer gains encourages a public role in southern pine research. Solidwood producers gain only minimally and pulpwood producers are gross losers as they pass more than their full gains on to consumers. This conclusion anticipates the more general observations that industrial landowners seem to be reducing their research budgets and relying relatively more heavily on NIPF producers. (NIPF lands produce approximately 70 percent of annual southern pine harvests. Industrial lands produce approximately 15 percent and public lands produce the remainder.) Our conclusion also anticipates that even those NIPF landowners with clear timber management objectives have little private incentive to encourage research in new timber management technologies.

Table 9.3. 3SLS estimates for southern pine pulpwood and solidwood markets

	pulpwood demand		pulpwood supply		solidwood demand		solidwood supply
intercept	731,700 (570,600)	intercept	-776,000* (181,100)	intercept	-1,645,000* (501,900)	intercept	888,900* (168,800)
own price	-9,092** (5,127)	own price	4,921** (19.53)	own price	-3,162** (1,423)	own price	3,072* (374.1)
final good price	1,800 (6,387)	inventory	0.02246* (0.001260)	final good price	21,730* (6,943)	inventory	0.00648* (0.00167)
labor	304,300* (97,570)	solidwood price	491.1** (239.8)	labor	114,900** (50,730)	pulp/paper price	-11,280* (1,694)
capital	-2,341,000* (764,900)			capital	1,025,000** (731,400)		
lagged production	0.2265** (0.1004)			lagged production	0.7860* (0.1299)		
R^2	.869		.952		.938		.934
Durbin-Watson	1.116		1.034				1.475
Durbin's h					0.821		
F	31.84*		171.89*		72.62*		122.65*
degrees of freedom	24		26		24		26

Numbers in parentheses are standard errors.
  * Significant at the 1 percent level
 ** Significant at the 5 percent level
*** Significant at the 10 percent level

Table 9.4. Gross benefits, 1950-1980, from southern pine research (millions of 1967$, undiscounted)

| | annual research-induced production shift | | | | | |
| | 0.5% | | | 1.0% | | |
benefitting group	consumers	producers	total	consumers	producers	total
market						
solidwood	18.44	2.10	20.54	36.86	4.25	41.11
pulpwood	31.58	-8.59	22.59	63.03	-16.86	46.18

A Closer Look at Distribution

The $40-80 million of undiscounted gross research benefits are transfers redistributed from national taxpayers to the regional factors of southern pine forestry and wood products production. The usual tests of redistributive programs have to do with impacts supporting disadvantaged populations. Therefore, the redistributive merit of southern pine research has to do with wage and employment gains for the unemployed and for low income southern workers. We used the US Forest Service's input-output (I-O) model (IMPLAN, Alward and Lofting 1990) to determine the direct, indirect, and induced impacts of public research on these populations.

The special features of our I-O analysis are its focus on (a) forward research impacts and (b) three industrial centers within the South, and (c) its temporal context, the 1982 base of IMPLAN and a 3-5 year cycle of macroeconomic expansion. (a) Research benefits are small for solidwood producers and negative for pulpwood producers. Furthermore, the labor component of timber production is very small. Therefore, the backward linkages of forestry research are unimportant from a distributive perspective. (b) Our industrial center are south Georgia and Bogalusa, Louisiana, two hubs of the forest industry, and coastal North Carolina, a locale with great potential gains from forestry research in drainage, site improvement and intensive management.

Finally, (c) I-O implies an assumption of constant technology in all unexamined sectors of the economy. Therefore, I-O is incompatible with long-term assessments. This forces us to seek an alternative to the 30-year period of our previous analyses. Our alternative is the 3-5 year period of a normal industrial expansion, a period short enough for the constant technology assumption of I-O to be meaningful yet long enough to measure significant impacts of research-induced increases in forest productivity. Specifically, our I-O assumption is a 3 percent timber productivity change. This is a generous equivalent to 3-5 years of the 0.5-1.0 percent annual southern pine research productivity change that we observed for the 1950-1980 period.

Table 9.5 reports the basic I-O results for the three regions. The base refers to 1982 levels in absence of the 3 percent research-induced productivity increase. Impacts are measured as changes in the wage bill (W), employment (N), and total industrial output (TIO) within six forest products industries and for the entire regional economy.

The results are unimpressive in all three regions and under two alternative solidwood and pulpwood allocations. Converting the table 9.5 results to relative terms, the most optimistic south Georgia impacts are for a 0.11 percent employment increase for the total economy, and a 1.1 percent increase in forest sector employment. The comparable impacts for Bogalusa are an 0.23 percent increase in total employment, and a 2.1 percent increase in forestry sector employment. The comparable impacts for coastal Carolina are an 0.06 percent increase in total employment, and a 1.7 percent increase in forestry sector employment. These small regional employment impacts are even less important when we consider that they are not immediate, but occur over a 3-5 year period. The comparable wage effects are never important in the six lower paid (of seventeen) sectors in any of the three regional economies.

The lack of substantial employment and wage effects in three regional economies with notable forestry sectors suggest no reason to anticipate more important employment and wage effects elsewhere in the South. In conclusion, the distributive impacts are unimportant in any scenario and the distributive merit of southern pine research is unlikely.

The Efficiency Benefits of Southern Pine Research

Determining measures of the efficiency gains from timber growth and management research requires knowledge of the R&D expenditures associated with the undiscounted gross benefits reported in table 9.4. Historic budget data are unavailable for many of the institutions involved in southern pine research. Therefore, we began with known US Forest Service budgets for the Southern and Southeastern Forest Experiment Stations. These must be reduced by those research expenditure categories not clearly and fully associated with southern pine growth and management (*e.g.*, forest recreation). We can extrapolate expenditure series for the other research participants—for state and university research, industry research, and forestry extension—from occasional direct evidence over the years, and from periodic references to the size of these expenditure categories relative to Forest Service expenditures. We disregard all research expenditures prior to 1935, the uncertain level of expenditures by university/industry coopera-

Table 9.5. Distributive impacts of southern pine research (in three select regions and over a 3-5 year period)

	south Georgia			Bogalusa environs			coastal Carolina		
	wage bill	employment	total industrial output	wage bill	employment	total industrial output	wage bill	employment	total industrial output
aggregate regional economy[a]									
base (1982)	6,416	420,365	21,014	1,913	130,785	6,450	3,166	214,207	10,118
with pine research increment	+6.9	+401	+26.9	+5.0	+307	+19.8	+2.05	+122	+8.6
percent change	.11	.10	.13	.26	.26	.31	.06	.06	.08
forest sector changes[b]									
logging	.87	47	4.39	.59	18	2.98	.28	14	1.41
sawmills	2.08	91	7.42	.89	52	3.19	.53	23	1.90
hardwood lumber	.01	1	.02	.00	0	.01	.00	0	.01
plywood	1.89	95	7.23	.81	52	3.17	.52	38	1.91
other wood products	.03	2	.14	.02	1	.07	.00	0	.01
pulp and paper	.02	1	.10	.64	22	2.77	.25	6	1.52

[a]Wage bill in millions of 1982$, employment in numbers of employees, total industrial output in millions of 1982$.
[b]Raw numbers, as shown in tables 7.4, 7.8, 7.12

tives, and the uncertain contribution of general forestry education to improving southern pine yields—all in the name of conservative total cost estimates. Conservative cost estimates, together with generous benefit estimates, will provide a confident upper bound for observations on the efficiency of timber management research.

Our final cost estimates suggest that the Forest Service was the only significant southern pine research institution before 1950. The other research institutions became important in the 1960s and 1970s as total research budgets rose to three times the Forest Service budgets. The total budget for all institutions continued rising until it was five to seven times the Forest Service research budget in 1980. Industry budgets decreased sharply in the 1970s but increasing state and university budgets offset these industry decreases.

The research-productivity lag poses an insurmountable estimation problem for timber growth and management. Our alternative is to examine a full range of hypothetical lags. We consider lags from zero to twenty years in five year increments and compare costs and benefits in all five cases. We begin with cost year cohorts representing accumulations of the annual expenditures that induce our measured 1950-1980 southern pine research benefits. The first cost cohort is 1935-1960, the second is 1935-1965, etc. The first cohort is the smallest research cost accumulation but it implies a maximum twenty-year lag before its last research expenditures (1960) begin to affect harvests (1980). Successive cost cohorts each add five years of estimated research expenditures, implying five additional years of costs but five fewer years in the maximum research-productivity lag. We also consider a range of possible social discount rates, and benefits arising from the range of research-induced inventory shifts between 0.5 and 1.0 percent annually.

Table 9.6 shows the results. Net benefits are negative for all but the single most extreme case. Even then they remain small at an accumulated $29 million for the entire thirty-year period of research impacts—or an average of only $80,000 per state per year. The second part of table 9.6 provides an additional check for the robustness of these results by comparing only the Forest Service share of all southern pine research costs with our total benefit estimates. Net benefits remain negative for the most reasonable cases of positive social discount rates and moderate research-productivity lags.

In conclusion, annual *physical* research gains of 0.6 percent may satisfy industrial researchers but the social efficiency of these gains is uninspiring at best for southern pine growth and management. This conclusion holds for the sum of a broad collection of research activities. It is important to recognize, however, that these results do not preclude the possibility that select and more specific research activities, perhaps various nursery-oriented research activities or weed control research, may display more satisfying results.[10] Nevertheless, because southern pine is generally considered the best example of timber management research success, it is probably unnecessary to examine other broad regional cases or forest types. We anticipate similar, or poorer, results for them.

Reflections on Timber Growth and Management Research

Can we believe this empirical observation of generally poor economic performance for southern pine management research? What explains it: perhaps low long-run timber production costs? Low production costs would leave few opportunities for cost-saving production gains, therefore little inducement for research and few final opportunities for introducing technological improvements.

Most timber production research has been of the land-saving or capital-(investment time)-saving variety. This is a reasonable outgrowth of what are often natural transfers from agriculture research. It is also reasonable because land and capital, not labor, are the predominant physical inputs. Land and capital have not been economically scarce inputs in forestry, however, and they will not become scarce inputs so long as there are stands

Table 9.6. Net benefits, 1950-1980, from southern pine research (millions of 1967$)

| social discount rate | present net value, with full estimated costs | | | | | |
| | 0% | | 4% | | 7% | |
research-induced production shift	0.5%	1.0%	0.5%	1.0%	0.5%	1.0%
cost cohort						
1935-1960	-15	29	-89	-13	-243	-121
1935-1965	-81	-38	-217	-141	-450	-328
1935-1970	-178	-135	-372	-296	-667	-546
1935-1975	-320	-276	-558	-482	-894	-773
1935-1980	-527	-483	-780	-705	-1,130	-1,008

| social discount rate | present net value, with US Forest Service Southern and Southeastern Forest Experiment Stations' research expenditures only | | | | | |
| | 0% | | 4% | | 7% | |
research-induced production shift	0.5%	1.0%	0.5%	1.0%	0.5%	1.0%
cost cohort						
1935-1960	16	60	-13	63	-96	26
1935-1965	-5	39	-53	23	-162	-40
1935-1970	-35	9	-101	-25	-229	-107
1935-1975	-67	-24	-144	-68	-281	-160
1935-1980	-102	-58	-181	-106	-321	-200

of old growth timber, abandoned old fields reverting to timber, and remaining timber harvest opportunities on less accessible sites.

This is the general case for southern pine. NIPF landowners are well-known for investing very little in their 70 percent of southern forests. Furthermore, where NIPF landowners do invest in forestry, their costs have been held down by a long history of federal cost-sharing programs and there is good reason to expect that these programs will continue. An additional large share of the current NIPF inventory of standing timber originally regenerated naturally and it still grows with little or no active management input (Binkley 1981, Royer 1985, Boyd and Hyde 1989, Holmes 1989, Hyberg 1989). Some is the free result of the forest conversion on old abandoned fields. Old field conversion may continue as tobacco declines in importance and old tobacco fields regenerate naturally.

These observations are consistent with our expectations from chapter 2 and with the frontier hypothesis for development of the forestry sector of the US economy. Forest managers plan harvests every year without also planning to introduce active plantation silvicultural practices on a large share of the harvested lands. Indeed, we observe that their plans are borne out. Even among southern industrial forests, less than 40 percent are actively managed plantations (USDA Forest Service 1988). Therefore, it should not be a surprising observation that current stumpage prices do not cover the costs of active silvicultural investments for all the acres from which timber is harvested. It is *often* not economically rational to invest in growing timber. Removing the bounty of naturally occurring timber, however, must eventually cause stumpage supplies to decrease. Stumpage prices will rise in anticipation of this decrease and will induce continued forestry operations and more active, more intensive forest management in the future.

The available empirical evidence suggests that this transition from frontier to developed forest economy has been going on for some time. Relative roundwood prices probably have been rising for over a century (Ruttan and Callaham 1962, Barnett and Morse 1963, Phelps 1975, Manthy 1978).[11] We might reasonably anticipate that they will continue to rise gradually until they finally equal the full long-run (timber rotation) costs of active silvicultural management on most forestlands. [There will always be some marginal lands contributing some share of all harvests from unmanaged, naturally occurring, low opportunity cost timber (Hyde 1981, Sedjo and Lyon 1990).]

This hypothesis of forest sector development explains why, today, the physical successes of timber growth and management research might be recognized at the experiment station but seldom implemented in the field. With the current level of standing natural inventory and current stumpage prices, the known technological improvements in timber management reduce costs an amount sufficient to pay for their implementation on only a few acres, perhaps only the 26 percent currently in pine plantations. Only 30 percent of this 26 percent is industrial. Technological improvements in southern pine, today, may be justified mostly on this subset of industrial lands. Historically, these industrial lands alone have been insufficient to support the full public and private southern pine research budget. Declining industrial research budgets suggest that the industry has drawn this conclusion for itself.

In sum, there has been little resource scarcity inducement to encourage research or to adopt technical change in southern pine. We anticipate that the same is true for those other forest types and other regions, like the Rocky Mountains and the Pacific Northwest, where a natural bounty of timber remains.[12]

Nevertheless, relative prices cannot continue increasing indefinitely and there is some expectation that the long-run upward price trend will deteriorate within the next ten to thirty years (Berck 1978, Libecap and Johnson 1978, Lyon 1981). It may have tapered off already in the South (Dutrow *et al.* 1982, Cubbage and Davis 1986). Prices will eventually attain the level of full active silvicultural production costs on more acres before they cease to increase any further (in the absence of unanticipated shifts in demand).

This means that the future may be more promising for southern pine research investments. It also means that we should advise care with respect to research budget cutbacks. The potentially long lags between timber growth and management research and its eventual impact on productivity suggest that research capacity today is a necessary prerequisite for satisfying anticipated demands for cost-reducing southern pine technologies in ten to thirty years. Nevertheless, the prior poor performance of southern pine research in the aggregate surely urges the utmost care in choosing which timber growth and management research budgets to continue supporting. Furthermore, it encourages greater care and greater reluctance for support of timber growth and management research for the other species and other forest types which probably have an even less distinguished history of research productivity than southern pine.

FINAL REFLECTIONS AND BROADER POLICY IMPLICATIONS

This final section reflects on the results of our empirical analyses and anticipates their meaning for the future of forestry research in the US, and comments on the continuing public role in forestry research. The section closes by drawing implications from the American experience for both the developed and the developing countries of the rest of the world. We speculate that available forest inventories and final consumer products are critical indicators of research policy implications different from those for the US.

Policy Implications for American Forestry Research

Our analyses support the public policy advice to concentrate forestry research on forest products and to fund timber growth and management research only most lightly and most carefully. Concerns for equity; including laborers, small landowners, and rural development; do not alter this recommendation.

Forest products research has an entirely more satisfying past than timber management research but a past that still encourages cautious decisions regarding both justifiable public roles and future budget allocations. Our analyses justify a public role in research for each forest products industry we examined. Careful review, however, shows that the specific justification varies from industry to industry and that there is no reason to anticipate that the general justification for a public role is universal across all forest products industries. Future investments in these and other forest products industries require prior inquiry into the nature of each industry's production costs, its research-productivity lag, the likelihood of research adoption, and the industry's market concentration, in order to determine the independent justification for a public research role in that industry.

Furthermore, successful research performance in the past is not a necessary indicator of outstanding future performance. For example, softwood plywood research was a great success between 1950 and 1980 but the SWPW industry has been partially replaced by the newer products of the structural particleboard industry. The growing market share of the latter industry (and declining share of the former) suggests a greater future potential for particleboard research.

Experience also recommends caution in adjusting research budgets to short-term political perceptions rather than longer-term market signals. For example, higher level government directives responding to both energy and environmental concerns in the 1970s failed to produce new cost-reducing technologies for the woodpulp industry. These directives did, however, divert Forest Products Laboratory research resources that had a previous history of more socially efficient production.

Nevertheless, our summary statement for forest products research is that previous research, on the whole, has been socially rewarding. We anticipate the same for the future, both in narrow and measurable economic efficiency terms and in broader social welfare terms. For example, we anticipate successful product-altering research, like some research in the woodpulp and wood preservatives industries which improves environmental quality. The benefits of product-altering environmental research, of course, fail to appear in those economic accounts of research and technical change that can only measure process innovation. We also anticipate that forest products research and new technologies save on wood utilization and, therefore, save standing timber and forestland for other, often non-market, land uses. US Forest Service calculations, for example, suggest that the truss frame construction technology may save a volume of wood greater than annual programmed harvests on all lands proposed for wilderness withdrawal.[13]

Indeed, the truss framing example nicely addresses another important question for forest policy. Forest land use, with competing industrial, recreational and environmental demands, is one of the most important natural resource issues of our time. Apparently, forest products research is a more elastic substitute for forest land than timber management research. Forest products research is also a better public investment than public timber management itself. Therefore, investing in forest products research may be one of the better ways to reduce the land use conflict and to save forests for competing non-timber uses.

This observation gains significance as we reflect on its implications for the important global issues of climate change and biological diversity. Forests provide the habitat for many endangered species and trees sequester the carbon that controls climate change. Apparently, forest products research, by decreasing the demand for wood and wood fiber, contributes more than timber management research to protecting the forest and these global values. We anticipate that biological forestry research may be better advised to concentrate on identifying critical species and habitat and improving their management.

Various other researchers find approximately 2 percent annual rates of technical change across the forest product industries. They also observe a labor-saving bias in technical change (Kendrick 1961, Stier 1980, Greber and White 1982, Buongiorno and Lu 1989). Therefore, the rates of land- and capital-saving change are somewhat less. Even land- and capital-saving rates of 1.5 percent, however, are more than double the 0.6 percent average annual rate we observe for southern pine. This means that, for a constant output of forest products, the uses of roundwood and capital facilities as inputs decline at the rate of 1.5 percent annually. It also means that, in 48 years, research and technical change will permit the forest product industries to consume just one-half their current consumption of all roundwood inputs while producing the same volume of final product. In contrast, it takes timber management research 120 years to double the volume of southern pine volume. It takes even longer for timber management research in other US species and it would take longer for southern pine if southern pine research investments were reduced to a lower, more socially efficient level.

In sum, forest products research has been a good investment. In some cases, it would have been an even better investment with additional funding. If carefully managed, it promises to be a continuing good research investment. It is a better use of scarce public funds than most timber management research. Furthermore, its past record suggests that it can be a key to reducing future demands for raw materials and, therefore, to saving forestland for other resource and environmental values. Finally, and regardless of the great potential for well-chosen research, the great variety in our analytical results argues most strongly for careful *ex ante* research analyses preceding specific commitments of public funds.

Implications for Forestry Research in the Rest of the World

These US experiences are only partially transferable. We must examine the specific market situations and policy incentives in other countries, for both forest products and timber growth and management, and contrast them with US markets and incentives before we can determine which conclusions transfer from the US forestry research experience to the rest of the world. We might anticipate that our general encouragement of forest products research does transfer, with emphasis on products with large markets. General conclusions for timber management research are more difficult. There may be four classes of cases, two in developed countries and two in developing countries, depending on local market and incentive conditions for timber and forestland in each country.

Developed countries: Forest products research opportunities for developed countries are comparable to those for the US. The products are similar and the markets are generally well-developed. Price-induced research and technical change may generally pay off at a high rate of social return, much as it does in the US.

The timber inventory and forest land situations may distinguish two developed country alternatives for timber management research. Canada and the Soviet Union are countries like the US, countries with large extensive margins of timberland. Indeed, Canadian and Soviet forest inventories are probably subject to a lower average level of active silvicultural management than US forest inventories. This available resource holds down stumpage values in Canada and the Soviet Union. Therefore, we expect that land and timber are relatively less scarce factors of production than labor and capital facilities, that the economy-wide incentives inducing labor- and capital-saving research are greater than the incentives inducing land- and timber-saving research, and that technical change will proceed only at a slow rate in the timber management industry. Forestland development follows the frontier hypothesis of economic growth. Like the US, there may be select and scattered timber management research opportunities in Canada and the Soviet Union, but general timber management research probably is not a highly productive activity in these countries.

This conclusion is consistent with Sedjo and Lyon's (1990) finding that, in an integrated world market, intensive forestry would be practiced on only the very best sites. Sedjo and Lyon find that, in the US South, only modest levels of investment in regeneration are optimal. In much of Canada, optimal management levels are low cost and very extensive.

Japan and the most of the developed countries of western Europe *may* suggest a different case for timber management research. These countries arguably possess no substantial extensive land margin supporting a natural bounty of timber and there are few old fields freely reverting to forests. In this case, the impact of wood and wood fiber imports from North America, the Soviet Union, and tropical countries with a forest frontier will determine the viability of Japanese and western European timber management research.

The impact of imports will be *important* for timber management research if imports are large enough to have an impact on domestic stumpage prices. This means that imports will have kept Japanese and western European stumpage from achieving a level of scarcity comparable to the relative scarcity of substitute resources. Therefore, the extensive timber frontier in other parts of the world is also an economically important frontier for Japan and western Europe. In this case, timber management research will not be a socially attractive investment, even in Japan and western Europe. It will become more attractive for Japan and western Europe as it becomes a better investment in the exporting countries.

Alternatively, the impact of wood and wood fiber imports will be less important for timber management research if relative Japanese and European stumpage prices are changing at a rate comparable with the rate for substitute resources in these economies. In this latter case, timber management would be a developed industry in Japan and western Europe and the price incentives for timber management research and technical change would be fully operative and comparable with those for the rest of the Japanese and western European economies. Timber management research would be fully as attractive as forest products research in this case.[14]

Determination of the correct alternative explanation for developed countries like Japan and those in western Europe must wait for (a) inference from assessments of regional relative stumpage price trends or (b) direct evidence from individual timber management research appraisals comparable to our assessment of southern pine research in the US.

Developing countries: There is less processing and the markets for processed wood and fiber products are less extensive in the developing countries than in the US. Therefore, the scope for forest products research in general is smaller in most developing countries than it is in the US. The result may be fewer forest product research opportunities, but those research opportunities that do exist probably offer the same high social rewards that forest products research offers in the US and other developed countries.

The logging industry may be a particularly good candidate for developing country forest products research. Our southern pine results and our extrapolations from them to other species and forest types and other parts of the world say nothing about logging research investments. Standing timber priced on the stump (stumpage) is our physical market measure in the southern pine case. Logging is the next higher step in the production process. Logging technologies are more labor- and capital equipment-intensive and less land- and resource-intensive than timber growing and management. Therefore, they are similar to forest products technologies as likely candidates for productive research investments.

Most modern logging technologies were developed in economically advanced countries where labor is scarcer relative to capital equipment. This relative difference in factor endowments may suggest opportunities in many developing countries for modern, yet relatively more labor-intensive, logging technologies. The absence of good roads and the range of topographic conditions across all developing country harvesting situations suggests further opportunity for either capital specialization or greater reliance on labor's versatility. In any case, logging technologies specially adapted to developing country conditions may be a rewarding forest products research opportunity.[15]

The market for wood as a domestic fuel for heating and cooking is an additional large market, therefore an additional forest products research opportunity, that is not so important in most developed countries. This market provides incentives for additional wood processing technologies that may be less important in developed countries. The obvious examples are charcoal research and research on improved stoves. The latter is a well-known recent success in many developing countries. Research to improve the efficiency of fuelwood consumption may have additional appeal from an equity perspective. The poorest households spend the largest household budget shares and have the highest income elasticities for fuelwood and its substitutes (Hyde and Amacher 1991).

Timber management research, as always, is more problematic. Many developing countries possess extensive stands of mature forest. The depths of the Amazon and Congo watersheds and the northeast of Thailand are examples. These forests compare with the US, Canadian, and Soviet case. A large extensive margin of forestland holds timber production costs low. Timber is not scarce relative to other factors of production and there is little economic incentive for timber management research. Timber management research in this case will have little social payoff.

Other developing countries may have less extensive forest resources but insecure tenurial rights for their remaining timber resources. The arid lands of India, the nationalized forests of Nepal, and the uplands of the Philippines are examples (Hyde and Amacher 1991). Insecure tenure encourages immediate local resource-consuming self-interests and removes any incentive for long-term resource management. Therefore, it dissipates all economic incentives associated with the basic timber and forestland resources and causes resource depletion. In this case, there is no price incentive for timber management research and research on timber management will be a misuse of public funds until the tenure problem is solved.

Finally, there are numerous cases of unintended but perverse policy spillovers to the forest resource. Binswanger (1989), for example, identifies a long list of macroeconomic policies and policies directed at rewarding other sectors; e.g., agriculture, livestock, mining; that encourage Amazonia's conversion to non-forest uses. Boyd et al. (1991) measures the impacts of a similar list for the Philippines. These policy spillovers come to rest on the forest resource because forests are generally a residual land use often without a specialized market-oriented constituency. These policies are equivalent to downward relative timber price shifts. They act as further disincentives for timber management research.

These three—a large margin of extensive timber management, insecure resource tenure, and perverse policy spillovers—are important cases in very many developing countries. They cause dissimilarity between forestry and the conditions which have induced the successful agriculture research experiences that are well-known for many developing countries. Therefore, direct comparisons with agriculture research in these countries, or with timber management research in developed countries, only conceal the important production relationships and the real policy issues. Unreasoned arguments to the contrary are distressing.

It seems to us that the altogether different issues of insecure resource tenure and perverse policy spillovers are the *necessary* first issues of policy inquiry in these cases. The products of timber management research, whether the timber is for domestic or for market consumption, cannot be introduced successfully until these issues are dealt with better. Tenure and perverse policy spillovers from non-forestry sectors are the better initial target for public research investments.

This said, we must also recognize another, more satisfying, timber management research case in many developing countries. Tenure is secure on most agricultural land and we observe many farmers who recognize the personal benefits of tree crops. We also observe some successful commercial forest plantations. Economic incentives are fully operative for the subset of these tree crops and these commercial plantation species for which the natural forests on the extensive margin of land with insecure tenure are unsatisfactory substitutes. This is a small but increasing share of developing country timber resources. The economic incentive exists, but the small overall size of these farm forests and commercial plantations suggests generally small returns to timber management research.

This recommends timber management research concentrating on those few species with large and widespread markets, as well as minimum competition from species growing naturally on the extensive margin. This recommendation is consistent with the more biological research orientations of Buckman (1988) and Davis *et al.* (1988). Buckman also encourages forestry extension in these cases. Successful extension improves research benefits by decreasing the communication time between successful research and its implementation by the many small forest farmers.

As the natural forests are depleted, as institutions adjust to ensure secure tenure, and as policymakers reconsider perverse policy spillovers, then we can anticipate that this final case will become more characteristic of developing country timber management. Farm forestry and commercial plantation forestry will become more prominent and the economic and social incentives for timber management research will increase. This may, however, take some time in most developing countries. Our evidence argues that the time has not yet arrived in the southern pine region of the US. It probably has not yet arrived elsewhere in the US, in Canada, or in the Soviet Union. Therefore, we must urge the greatest care in choosing timber management research activities, particularly as a widespread use of public funds in developing countries.

CONCLUSION

Forestry research has a mixed performance history in the US. It promises mixed opportunities, in the US and worldwide, in the future. A public role is often justified, particularly in forest products research. A public role is more doubtful in timber management research. Distributive concerns for rural development and disadvantaged local populations do not alter this conclusion.

These mixed results point to the need for the careful applications of analytical logic and quantified empirical evidence. Short-term public perceptions and political influence are often in error and inappropriate comparisons of agriculture research experience with timber management research are easy. Non-measured product-quality research benefits can be important—but they should be segregated from measurable cost-reducing research benefits and submitted to careful, if necessarily subjective, judgment.

In sum, forestry research has a great but selective promise. The public is probably underinvested in forest products research, in the US and worldwide. Forest products research in general does follow the historic pattern of high payoffs in agricultural research, although returns to forest products research are more variable than returns to agriculture research. Forest products research promises large future gains in social efficiency, as well as additional resource gains as it provides substitutes for timber harvests from lands with competing, non-timber uses.

Indeed, forest products research, through its demand-reducing impacts on the timber input, has a greater impact than forest management research on timber availability and forest land use. This is a particularly important observation in the presence of global concerns with deforestation. This conclusion provides an additional non-market (climate change and biodiversity) justification for forest products research and it frees biological forestry researchers to concentrate on identifying critical forest species, habitat, and their management alternatives.

The public is overinvested in timber management research in the US and, perhaps with select exceptions, the American public would be well-served with a cautious reduction in timber management research investment. This observation is also probably accurate for Canada and the Soviet Union, and perhaps for Japan and western Europe and for most developing countries. Timber management research investments in developed countries may become better investments as the stocks of extensively managed timber are drawn down. For the US South, this may mean in the next ten to thirty years. Timber management research investments in developing countries will become better investments as their extensively managed stocks are drawn down, as their institutions adjust to provide more secure tenure for their land and tree resources, and as their policymakers revise policies with perverse secondary impacts on forest resources. Until then, these latter social science issues will provide a better focus for forestry research in many developing countries.

The historical range of forestry research results, the technical difficulties of accurately predicting future research payoffs and the unreliability of political judgment about wise research ventures, altogether suggest that the best public investments may be in human capital. Training and employing scientists and engineers with broad skills and with abilities to respond to market incentives for new forest products and production processes may be the best investments of public research funds.

NOTES

1. Mergen *et al.* (1988) provide a global estimate of forestry research expenditures. They estimate that global forestry research expenditures approximately doubled in ten years to $1.024 billion (1980$) in 1980. Expenditures in developing countries alone increased from $79 million to $198 million. More recently, the Consultative Group of International Agriculture Research institutions agreed in December 1988 to extend their agricultural activities by introducing three new forestry research networks, one each in Asia, Africa, and Latin America.

2. Disembodied technical change is factor neutral. It includes all technical change that occurs without attribution to an identified technical production input. Disembodied technical change is usually explained econometrically as a function of some independent time variable.

3. Boyd and Hyde (1989) provide some context for forestry. They chance the estimate that the static aggregate welfare losses from various regulations in the forestry sector of the US economy (*e.g.*, trade restrictions, public ownership, preferential taxation, environmental harvest restrictions, timber production incentives, etc.) may exceed 20 percent. The largest of these, preferential capital gains taxation, was eliminated in the 1986 federal tax law. Therefore, static welfare losses for the entire forestry sector of the US economy are surely less than the Boyd-Hyde 20 percent estimate today. Forestry research results comparable to those in agriculture would quickly offset the revised Boyd-Hyde estimate of static losses—and quickly make these static constraints on productivity and social welfare seem small and unimportant.

4. The patent system intends to protect against this third class of cases. It fails when the new technology can be modified readily or when alternative designs can create similar production technologies.

5. That is, an independent time variable in the production/supply equation is statistically insignificant.

6. Our estimation process for specifying this lag proceeds in three steps: (1) Obtain expert opinion: FPL scientists estimated that there was a two year lag before

the initial R&D impact on productivity. (2) Obtain two-stage least-squares estimates for an index of SWPW technology as functions of public and private R&D each lagged from one to seven years: The statistically most satisfying estimate is for two years. (3) Use these two year research-productivity lags in the NL2SLS demand and supply estimates (subscripts on public and private R&D in table 9.1), but retest the final demand-supply specification for alternate R&D lags: Alternate R&D lags do not improve on the SWPW demand-supply estimates in table 9.1.

7. There were approximately eighty firms in SIC 2436 in the late 1970s. The four largest shared approximately 38 percent of the market in 1978.

8. There is an argument that Swedish sawmill research produced a substantial positive spillover to American sawmills. If this spillover were statistically significant, then it would cause large error terms in our equations and an unsatisfactory coefficient of determination. Yet the coefficient of determination is an acceptable 0.67. Furthermore, if Swedish research productivity had a regular effect over time, then it would argue for a positive-signed term for disembodied technical change. Yet preliminary supply regressions with terms for disembodied technical change were less satisfactory in general and the coefficient on disembodied technical change both had the wrong sign and was statistically insignificant. Thus, we reject the argument for Swedish, or other international, research impacts on the US sawmill industry.

9. The interesting observation that pulpwood is a substitute in solidwood supply, yet solidwood is a complement in pulpwood supply, supports an earlier observation for Sweden by Johansson and Lofgren (1985).

10. Consider the economic literature on gains from nursery management and improved seed stock (Davis 1967, Porterfield 1974, Hyde 1981, Westgate 1986, Williams and deSteiguer 1990). Much of it anticipates that microeconomic speculations can transmit into aggregate, sector-wide gains, particularly in the southern pine region. This transmission may not hold without sharply declining stumpage price effects and resulting negative net revenues. The potential of herbaceous weed control research is better documented (Huang and Teeter 1990, Warren 1990). In sum, the literature raises the possibility of economic gains from some timber growth and management research.

11. This experience of rising relative prices for such a long period of time is unique for lumber, and perhaps timber, among all primary resources (Barnett and Morse 1963, Manthy 1978).

12. In the Rockies and the Pacific Northwest today, the naturally occurring timber is often the original bounty of an old growth forest which was present when the first white settlers arrived. It is seldom the result of economic adjustment and old field conversion.

This leaves the question of research investments and technical change in forestry in other regions of the US; e.g., the Lake States and the Northeast. Research success and technical change has been less frequent in these latter regions because these regions are less price competitive in today's national markets than either the South or the Pacific Northwest.

13. Specifically, the 28.6 million acres proposed for wilderness withdrawal in RARE II. (US Forest Service calculations provided by H.G. Wahlgren, February 15, 1989.)

14. Japan's log imports from North America, the Soviet Union, and the tropics were 62 percent of total domestic consumption in 1987. Japan's imports of roundwood equivalents from the same regions were 71 percent of total domestic consumption in 1987. The comparable western European import proportion from the same regions was, perhaps, one-third as great (J. Vincent, Michigan State University, personal communication, March 29, 1990). These data suggest the hypothesis that Japan may fall in the first category where imports from regions with forest frontiers are deterrents to price-induced technical change and economically viable timber management research. They also suggest the hypothesis that western Europe falls in the second category, and timber management research may be economically viable. We emphasize, however, that these data are weak evidence and that the Japan and western European cases beg further inquiry.

15. J. Douglas of the World Bank alerted us to this point. Laarman *et al.* (1981) provides analysis and empirical evidence in the Philippines.

LITERATURE CITED

Abt, R.C. 1984. Regional production structure and factor demand in the U.S. lumber industry. *Forest Science* 33(1):164-173.

Adams, D.M., and R.W. Haynes. 1981. *The 1980 softwood timber assessment market model: structure, projections and policy simulation.* Forest Science Monograph 22. (Results revised 1985, personal communication, D. Adams)

Alward, G.S. 1987. A generalized system for regional input-output analysis of natural resource issues. Pp. 162-174. In *Symposium on systems analysis in forest resources*, ed. P.E. Dress and R.C. Field. Athens: Georgia Center for Continuing Education.

Alward, G.S., and E.M. Lofting. 1990. *Draft—IMPLAN version 2.0.* USDA Forest Service Rocky Mountain Forest and Range Experiment Station. General Technical Report.

Alward, G.S., and C.J. Palmer. 1983. IMPLAN: an input-output analysis system for forest service planning. Pp. 131-140. In *Forest sector models*, ed. R. Seppala, C. Row, and A. Morgan. Oxford: AB Academic Publishers.

Anderson, R.G. 1982. *Regional production and distribution patterns of the softwood plywood industry.* Tacoma, Wash: American Plywood Association.

Baechler, R.H., and L.R. Gjovik. 1986. Looking back at 75 years of research in wood preservation at the U.S. Forest Products Laboratory. Pp. 133-149. In *Proceedings of the 82d annual meeting of the American Wood-Preservers' Association.*

Ballard, C.L., D. Fullerton, J.B. Shoven and J. Whalley. 1985. *A general equilibrium model for tax policy evaluation.* Chicago: University of Chicago Press.

Bare, B., and R. Loveless. 1985. A case history of the regional forest nutrition research project: investments, results, and applications. Report submitted to the USDA Forest Service North Central Forest Experiment Station under project number PNW 82-248.

Barnett, H., and C. Morse. 1963. *Scarcity and growth.* Baltimore: Johns Hopkins University Press for Resources for the Future.

Bengston, D.N. 1983. Forestry research evaluation: an example. Pp. 53-67. In *Economic evaluation of investments in forestry economics*, ed. W.F. Hyde. Durham, NC: Acorn Press.

Bengston, D.N. 1984. Economic impacts of structural particleboard research. *Forest*

Science 30(3):685-697.

Bengston, D.N. 1985. Aggregate returns to lumber and wood products research: an index number approach. Pp. 62-68. In *Forestry research evaluation: current progress, future directions*, ed. C. Risbrudt and P. Jakes. St. Paul, MN: USDA Forest Service, North Central Forest Experiment Station, General Technical Report NC-104.

Bengston, D.N., and H.M. Gregersen. 1986. Forestry research and income redistribution. Pp. 117-122. In *Proceedings IUFRO: evaluation and planning of forestry research, S6.06-S6.06.01*, comp. D.P. Burns. Upper Darby, PA: USDA Forest Service Northeast Forest Experiment Station General Technical Bulletin NE-111.

Berck, P. 1978. The economics of timber: a renewable resource in the long run. *Bell Journal of Economics* 9(3):447-462.

Binkley, C.S. 1981. *Timber supply from nonindustrial forests: a microeconomic analysis of landowner behavior*. New Haven, CT: Yale University School of Forestry and Environmental Studies Bulletin No. 22.

Binkley, C.S. 1985. Long run timber supply: price elasticity, inventory elasticity, and the capital-output ratio. Laxenberg, Austria: International Institute for Applied Systems Analysis working paper 85-10.

Binswanger, H. 1989. *Fiscal and legal incentives with environmental effects on the Brazilian Amazon*. Washington: World Bank discussion paper 69.

de Borger, B., and J. Buongiorno. 1985. Productivity growth in the paper and paperboard industries: a variable cost function approach. *Canadian Journal of Forest Research* 15(5):1013-1021.

Boyce, S.G., and H.A. Knight. 1979. *Prospective ingrowth of southern pine beyond 1980*. Asheville, NC: USDA Forest Service Research Paper SE-200.

Boyd, R.G. 1985. Government support of nonindustrial production: the case of private forests. *Southern Economics Journal* 51(1):89-106.

Boyd, R.G., and W.F. Hyde. 1990. *Forestry sector intervention: the impacts of public regulation on social welfare*. Ames: Iowa State University Press.

Boyd, R.G., and D.H. Newman, 1991. Tax reform and land-using sectors of the US economy. *American Journal of Agricultural Economics* 73(2):398-409.

Boyd, R.G., and B.J. Seldon. 1991. Revenue and land use effects of proposed changes in sin taxes. *Land Economics* 67(3):365-374.

Boyd, R.G., W. Cruz, and W.F. Hyde. 1991. Unintended public policy spillovers on upland resources and the environment. Ohio University Economics Department working paper.

Brannlund, R., M. Goransson, and K.G. Lofgren. 1985b. The effect on the short-run supply of wood from subsidized regeneration measures: an econometric analysis. *Canadian Journal of Forest Research* 15(5):941-948.

Brannlund, R., P.O. Johansson, and K.G. Lofgren. 1985a. An econometric analysis of aggregate sawtimber and pulpwood supply in Sweden. *Forest Science* 31(3):595-606.

Brumm, H.J., and J.M. Hemphill. 1976. The role of government in the allocation of resources to technical innovation. Washington: National Science Foundation 2:41-47.

Bruner, A.D., and J.K. Strauss. 1987. The social returns to public R&D in the U.S. wood preserving industry: 1950-1980. Research Triangle, NC: Southeastern Center for Forest Economics Research working paper no. 35.

Buckman, R.E. 1988. As summarized in Buckman on tropical forestry research. *ISTF News* 9(4):3.

Bullard, S.H., and T.J. Straka. 1986. Role of company sales in funding research and development by major U.S. paper companies. *Forest Science* 32(4):936-943.

Buongiorno, J., and H-C. Lu. 1989. Effects of costs, demand and lumber productivity on the prices of forest products in the United States, 1958-1984. *Forest Science* 25(2):349-363.

Buongiorno, J., and R.A. Oliviera. 1977. Growth of the particleboard share of production of wood-based panels in industrialized countries. *Canadian Journal of Forest Research* 7(2):383-391.

Callaham, R.Z. (technical coordinator). 1981. *Criteria for deciding about forestry research programs.* Washington: USDA Forest Service General Technical Report WO-29.

Campbell, R.C., and J.H. Hughes. 1980. Forest management systems in North Carolina pocosins. Pp. 199-214. In *Pocosin wetlands,* ed. C.J. Richardson. Stroudsburg, PA: Hutchinson Ross Publishing Co.

Chang, S.J. 1986. The economics of optimal stand growth and yield information gathering. Report submitted to the USDA Forest Service North Central Experiment Station under Cooperative Research Agreement 23-83-27.

Cline, P.L. 1975. Sources of productivity change in United States agriculture. Ph.D. diss. Oklahoma State University.

Connaughton, K.P., and W. McKillop. 1979. Estimation of "small-area" multipliers for the wood processing sector—an econometric approach. *Forest Science* 25(1):7-20.

Cubbage, F.W., and J.W. Davis. 1986. Historical and regional stumpage price trends. *Forest Products Journal* 36(9):33-39.

Davis, J. 1981. A comparison of procedures for estimating returns to research using production functions. *Australian Journal of Agricultural Economics* 25(1):60-72.

Davis, J., D. McKenney, and J. Turnbull. 1988. Potential gains from forestry research and a comparison with agricultural commodities. Draft manuscript. Canberra: Australian Centre for International Agricultural Research ACIAR/ISNAR project paper 15.

Davis, L.S. 1967. Investments in loblolly pine clonal seed orchards. *Journal of Forestry* 65(6):882-887.

Denison, E. 1974. *Accounting for United States economic growth, 1929-1969.* Washington, DC: Brookings Institution pp. 131-137.

Dervis, K., J. deMelo, and S. Robinson. 1982. *General equilibrium models for development policy.* Cambridge: Cambridge University Press.

Dutrow, G., J.M. Vasievich, and M. Conklin. 1982. Economic opportunities for increasing timber supplies. Pp. 246-254. In *An Assessment of the Timber Situation in the United States, 1952-2030.* USDA Forest Service, Forest Resource Report No 23.

Elrod, R.H., K.M. El Sheshai, and W.A. Schaffer. 1972. *Interindustry study of forestry sectors for the Georgia economy.* Athens: Georgia Forest Research Council Report No. 31.

Evenson, R.E. 1981. Research evaluation: policy interests and the state of the art. Pp. 196-212. In *Evaluation of agricultural research,* ed. G.W. Norton, W.L. Fishel, A.A. Paulsen, and W.B. Sundquist. St. Paul: Minnesota Agricultural Experiment Station Misc. Publ. 8-1981.

Fleischer, H.O., and J.F. Lutz. 1962. Technical considerations in the manufacture of southern pine plywood. Mimeograph, US Forest Products Laboratory.

Flick, W.A. 1986. Regional aspects of the Southern Timber Study. Auburn, AL: Auburn University School of Forestry (unpublished report).

Flick, W.A., P. Trenchi III, and J.R. Bowers. 1980. Regional analysis of forest industries. *Forest Science* 26(3):548-560.

Gilless, J.K., and J. Buongiorno. 1987. *PAPYRUS: a model of the North American pulp and paper industry.* Forest Science Monograph 28.

Grabowski, H.G. 1970. Demand shifting, optimal firm growth, and rule-of-thumb decision making. *Quarterly Journal of Economics* 84(2):217-235.

Gray, J. 1985. *Growth of southern higher education forestry programs and their impact on the South's timber resource and industries.* Washington: USDA Forest Service misc. publ. no. 1456.

Greber, B.J., and D.E. White. 1982. Technical change and productivity growth in the lumber and wood products industry. *Forest Science* 28(1):135-147.

Greber, B.J., and H.W. Wisdom. 1985. A timber market model for analyzing roundwood product interdependencies. *Forest Science* 31(1):164—179.

Gregory, G.R. 1966. Estimating wood consumption with particular reference to the effects of income and wood availability. *Forest Science* 12(1):104-117.

Griliches, Z. 1958. Research costs and social returns: hybrid corn and related innovations. *Journal of Political Economy* 66:419-431.

Griliches, Z. 1963. The sources of measured productivity growth: United States agriculture, 1940-1960. *Journal of Political Economy* 71(2):331-346.

Griliches, Z. 1964. Research expenditures, education, and the aggregate agricultural production function. *American Economic Review* 54(6):961-974.

Griliches, Z. 1979. Issues in assessing the contribution of research and development to productivity growth. *Bell Journal of Economics* 10(1):92-116.

Griliches, Z., and J. Mairesse. 1984. Productivity and R&D at the firm level. Pp. 33-75. In *R&D, patents, and productivity,* ed. Z. Griliches. Chicago: University of Chicago Press.

Guthrie, J.A. 1972. *An economic analysis of the pulp and paper industry.* Pullman: Washington State University Press.

Haddock, D. 1982. Basing-point pricing: competitive vs. collusive theories. *American Economic Review* 72(3):289-306.

Haxby, T.S. 1984. Returns to public investments in forestry research and development for sawmills and planing mills. Master of Forestry project, Duke University, Durham, NC.

Hay, D.A., and D.J. Morris. 1979. *Industrial economics: theory and evidence.* Oxford: Oxford University Press.

Haygreen, J., H. Gregersen, I. Holland, and R. Stone. 1986. The economic impact of timber utilization research. *Forest Products Journal* 36(2):12-20.

Haynes, R.W., and D.M. Adams. 1985. Simulations of the effects of alternative assumptions on demand-supply determinants on the timber situation in the United States. Washington: USDA Forest Service resource economics report.

Hertel, T.W., and M.E. Tsigas. 1988. Tax policy and US agriculture: a general equilibrium approach. *American Journal of Agricultural Economics* 70(2):289-302.

Hertford, R., and A. Schmitz. 1977. Measuring economic returns to agricultural research. Pp. 148-167. In *Resource allocation and productivity in national and international research*, ed. T.M. Arndt, D.G. Dalrymple and V.W. Ruttan. Minneapolis: University of Minnesota Press.

Hodges, D., F. Cubbage, and P. Jakes. 1988. Trends and distributional consequences of forest management research in the United States. Unpublished manuscript School of Forest Resources University of Georgia.

Holley, D.L. 1970. Location of the softwood plywood and lumber industries: a regional programming analysis. *Land Economics* 46(2):127-137.

Holmes, T. 1989. A household production model of non-industrial private landowner behavior. Draft manuscript. USDA Forest Service Southeastern Forest Experiment Station.

Horvath, G.M. 1980. Lumber, pulp and paper. Pp. 158-174. In *The improvement of productivity: myths and realities*, ed. J.E. Ullman. New York: Preager Publishers.

Huang, Y.S., and L. Teeter. 1990. An economic evaluation of research on herbaceous weed control in southern pine plantations. *Forest Science* 36(2):313-329.

Humphrey, D.B., and J.R. Moroney. 1975. Substitution among capital, labor, and natural resource products in American manufacturing. *Journal of Political Economy* 83(1):57-82.

Hussey, J., ed. 1985. *Timber mart south yearbook—1984*. Cambridge: Data Resources Inc., McGraw Hill.

Hyberg, B.T., and D. Holthausen. 1989. The behavior of non-industrial private landowners. *Canadian Journal of Forest Research* 19(8):1014-1023.

Hyde, W.F. 1981. *Timber supply, land allocation and economic efficiency*. Baltimore: Johns Hopkins University Press for Resources for the Future.

Hyde, W.F., and G.S. Amacher, eds. 1991. *Forestry and rural community development: an empirical examination from South and Southeast Asia*. New Delhi: Oxford and IBC Press.

Johansson, P.O., and K.G. Lofgren. 1985. *The economics of forestry and natural resources*. Oxford: Basil Blackwell.

Jones, K.D., and G.W. Zinn. 1986. *Forests and the West Virginia economy*. Morgantown: West Virginia Forest Development Service Agriculture and Forest Experiment Station.

Josephson, H.R. 1989. *A history of forestry research in the South*. Washington: USDA Forest Service misc. publ. no. 1462.

Kamien, M.I., and N.L. Schwartz. 1982. *Market structure and innovation*. Cambridge: Cambridge University Press.

Kaufert, F., and W. Cummings. 1955. *Forestry and related research in North America*.

Washington: Society of American Foresters.

Kendrick, J.W. 1961. *Productivity trends in the United States*. Princeton: Princeton University Press.

Kendrick, J.W., and E.S. Grossman. 1980. *Productivity trends in the United States: trends and cycles*. Baltimore: Johns Hopkins University Press.

Knutson, M., and L.G. Tweeten. 1979. Toward an optimal rate of growth in agricultural production research and extension. *American Journal of Agricultural Economics* 61(1):70-76.

Kuuluvainen, J. 1986. An econometric analysis of the sawlog market in Finland. *Journal of World Forest Resource Management* 2(1):1-19.

Laarman, J., K. Virtanen, and M. Jurvelius. 1981. *Choice of technology in forestry*. Quezon City, Philippines: New Day Publishers (for International Labor Office).

Lewandrowski, J.K. 1989. A regional model of the US softwood lumber industry. Draft manuscript, Agricultural Economics Department, North Carolina State University, Raleigh, NC.

Libecap, G.D., and R.N. Johnson. 1978. Property rights, nineteenth-century federal timber policy and the conservation movements. *Journal of Economic History* 39(1):129-142.

Lindner R.K., and F.G. Jarrett. 1978. Supply shifts and the size of research benefits. *American Journal of Agricultural Economics* 60(1):48-58.

Lyon, K.S. 1981. Mining of the forest and the time path of the price of timber. *Journal of Environmental Economics and Management* 8(4):330-345.

Mansfield, E. 1968. *The economics of technological change*. New York: Norton.

Mansfield, E. 1984. R&D and innovation: some empirical findings. Pp. 127-154. In *R&D, patents, and productivity*, ed. Z. Griliches. Chicago: University of Chicago Press.

Manthy, R.S. 1978. *Natural resource commodities—a century of statistics*. Baltimore: Johns Hopkins University Press for Resources for the Future.

McKeever, D.B. 1977. *Woodpulp mills in the United States in 1974*. Madison, WI: USDA Forest Service Resource Report FPL-1.

McKeever, D.B. 1987. *The United Stated woodpulp industry*. Madison, WI: USDA Forest Service Resource Bulletin FPL-RB-18.

McKillop, W., T.W. Stuart, and P.J. Geissler. 1980. Competition between wood products and substitute structural products: an econometric analysis. *Forest Science* 26(1):134-148.

Mergen, F., R.E. Evenson, M.A. Judd, and J. Putnam. 1988. Forestry research: a provisional global inventory. *Economic Development and Cultural Change* 36(1):149-171.

Merrifield, D.E., and R.W. Haynes. 1983. Production function analysis and market adjustments: an application to the Pacific Northwest forest products industry. *Forest Science* 29(4):813-822.

Mills, T.J., and R.S. Manthy. 1974. *An econometric analysis of market factors determining supply and demand for softwood lumber*. East Lansing, MI: Mich State Univ Agric Exper Sta Rep 238.

National Science Foundation. 1981. *Research and development in industry*. Washington: National Science Foundation Pub NSF 82-317.

Nautiyal, J.C., and L. Couto. 1981. The use of production function analysis in forest management: eucalyptus in Brazil, a case study. *Canadian Journal of Forest Research* 12(3):452-458.

Nautiyal, J.C., and B.K. Singh. 1985. Production structure and derived demand for factor inputs in the Canadian lumber industry. *Forest Science* 31(5):871-881.

Nerlove, M., and K.J. Arrow. 1962. Optimal advertising policy under dynamic conditions. *Economica* 29(1):129-142.

Newman, D.H. 1986. An econometric analysis of aggregate gains from technical change in southern softwood forestry. Ph.D. diss. Duke University, Durham, NC.

Newman, D.H. 1988. *The optimal forest rotation: a discussion and annotated bibliography.* Asheville: USDA Forest Service General Technical Report SE-48.

Newman, D.H. 1991. Changes in southern softwood productivity: a modified production function analysis. *Canadian Journal of Forest Resources* 21:1278-1287.

Norton, G.W., and J.S. Davis. 1981. Evaluating returns to agricultural research: a review. *American Journal of Agricultural Economics* 63(4):685-99.

Oster, S.M., and J.M. Quigley. 1977. Regulatory barriers to the diffusion of innovation: some evidence from building codes. *Bell Journal of Economics* 8(2):361-377.

Otto, D.M. 1981. An economic assessment of research and extension investments in corn, wheat, soybeans, and sorghum. Blacksburg, VA: VPI&SU Agric Econ Dep Sp-81-8.

Otto, D.M., and J. Havlicek, Jr. Undated. An economic assessment of corn, soybean, and wheat research and extension investments in the south and north central regions. CSRS/USDA mimeo to IR-6.

Pakes, A., and M. Shankerman. 1984. The rate of obsolescence in patents, research gestation lags, and the private rate of return to research resources. Pp. 73-88. In *R&D, patents, and productivity,* ed. Z. Griliches. Chicago: University of Chicago Press.

Palmer, C., E. Siverts, and J. Sullivan. 1985. IMPLAN analysis guide. Ft. Collins, CO: USDA Forest Service Rocky Mountain Forest and Range Experiment Station (unpublished report of the Land Management Systems Section).

Percy, M., and L. Constantino. 1987. A policy simulation model for the forest sectors of British Columbia and Canada. Pp. 181-192. In *Forest sector and trade models,* ed. P.A. Cardellichio, D.M. Adams, and R.W. Haynes. Seattle: University of Washington College of Forest Resources.

Peterson, W.L. 1967. Returns to poultry research in the United States. *Journal of Farm Economics* 49(3):656-669.

Phelps, R.B. 1977. *The demand and price situation for forest products, 1976-77.* Washington: USDA Forest Service Misc. Pub. No. 1357.

Porterfield, R.L. 1974. *Predicted and potential gains from tree improvement programs: a goal programming analysis of program efficiency.* Technical report #52. Raleigh: North Carolina State University School of Forest Resources.

Porterfield, R.L., T.R. Terfehr, and J.E. Moak. 1978. *Forestry and the Mississippi economy.* Starkville: Mississippi Agriculture and Forestry Experiment Station

Bulletin 869.

Repetto, R. 1988. Subsidized timber sales from national forests in the United States. Pp. 353-84. In *Public policies and the misuse of forest resources*, ed. R. Repetto and M. Gillis. Cambridge: Cambridge University Press.

Risbrudt, C.D. 1979. Past and future technological change in the U.S. forest industries. Ph.D diss. Michigan State University, East Lansing, MI.

Robinson, V.L. 1974. An econometric model of softwood lumber and stumpage markets, 1947-67. *Forest Science* 20(2):171-180.

Robinson, V.L. 1975. An estimate of technological progress in the lumber and wood products industry. *Forest Science* 21(2):149-154.

Rockel, M.L., and J. Buongiorno. 1982. Derived demand for wood and other inputs in residential construction: a cost function approach. *Forest Science* 28(2):207-219.

Rogers, E.M. 1983. *Diffusion of innovations*, 3rd ed. New York: The Free Press.

Royer, J. 1985. Market and policy influences on the reforestation behavior of southern landowners. Research Triangle Park, NC: Southeastern Center for Forest Economics Research working paper no. 12.

Ruttan, V.W. 1980. Bureaucratic productivity: the case of agricultural research. *Public Choice* 35(3):529-547.

Ruttan, V.W. 1982. *Agricultural research policy*. Minneapolis: University of Minnesota Press.

Ruttan, V.W., and J.C. Callahan. 1962. Resource input and output growth: comparisons between agriculture and forestry. *Forest Science* 8(1):68-82.

Sallyards, M.D. 1985. Returns to factors of production in the US pulp industry. M.A. paper, Ohio University Department of Economics, Athens, OH.

Schaffer, W.A., E.A. Laurent, and E.M. Sutter. 1972. Using the Georgia economic model. Atlanta: Georgia Institute of Technology.

Schallau, C., W. Maki, and J. Beuter. 1969. Economic impact projections of alternative levels of timber production in the Douglas-fir region. *Annals of Regional Science* 3(1):96-106.

Scherer, F.M. 1980. *Industrial market structure and economic performance*, 2nd ed. Chicago: Rand McNally.

Schmitz, A., and D. Seckler. 1970. Mechanized agriculture and social welfare: the case of the tomato harvester. *American Journal of Agricultural Economics* 52(3):569-577.

Schumpeter, J.A. 1950. *Capitalism, socialism and democracy*. New York: Harper.

Sedjo, R., and K. Lyon. 1990. *The long-term adequacy of world timber supply*. Washington: Resources for the Future.

Seldon, B.J. 1985. A nonresidual approach to the measurement of social returns to research with application to the softwood plywood industry. Ph.D. diss. Duke University, Durham, NC.

Seldon, B.J. 1988. R&D allocation in the competitive industry. Athens: Ohio University Economics Department.

Solow, R.M. 1957. Technical change and the aggregate production function. *Review of Economics and Statistics* 39(3):312-320.

Sonka, S.T., and D.I. Padberg. 1979. *Estimation of an academic research and development price index.* Univ of Illinois, Agricultural Economics Bulletin 79 E-1.

Spelter, H. 1985. A product diffusion approach to modeling softwood lumber demand. *Forest Science* 31(3):685-700.

Stier, J.C. 1980. Estimating the production technology in the U.S. forest products industries. *Forest Science* 26(3):471-482.

Styrman, M., and S. Wibe. 1986. Wood resources and availability as determinants of the supply of timber. *Canadian Journal of Forest Research* 16(2):256-259.

Sullivan, J. 1977. A review of forest and rangeland research. Pp. 38-44. In *A review of forest and rangeland research policies in the United States.* Washington: Renewable Natural Resources Foundation.

Teeter, L., G.S. Alward, and W.A. Flick. 1989. Interregional impacts of forest-based economic activity. *Forest Science* 35(3):515-531.

Troutman, F.H., and S.G. Breshears. 1981. *Forests and the Arkansas economy.* Little Rock: Industrial Research and Experiment Center Publication No. D-17.

Ulrich, A.H. 1983. *U.S. timber production, trade, consumption, and price statistics 1950-1981.* Washington: USDA Forest Service Misc Publ 1424.

Ulrich, A.H. 1985. *U.S. timber production, trade, consumption, and price statistics: 1950-1984.* Washington: USDA Forest Service, Misc Publ 1450.

USDA Forest Service. 1920. *Timber depletion, lumber prices, lumber exports and concentration of timber ownership.* Washington: USGPO.

USDA Forest Service. 1982. *An analysis of the timber situation in the United States, 1952-2030.* Washington: Forest Resource Report No. 22.

USDA Forest Service. 1982. *An assessment of the timber situation in the United States: 1952-2030.* Washington: Forest Resource Report No. 23.

USDA Forest Service. 1988. *The South's fourth forest: alternatives for the future.* Washington: Forest Resource Report No. 24.

US Department of Commerce. Bureau of Census. 1975. *Historical statistics of the United States, colonial times to 1970, bicentennial edition (part 2).* Washington: US Bureau of the Census.

US Department of Commerce. Bureau of Census. 1983. *Statistical Abstract of the United States: 1984.* Washington: USGPO.

US Department of Commerce. Bureau of Census. 1985a. *Census of manufactures, industry series: logging camps, sawmills, and planing mills.* Washington: USGPO document no. MC82-I-24A.

US Department of Commerce. Bureau of Census. 1985b. *Census of manufactures, industry series: pulp, paper and board mills.* Washington: USGPO document no. MC82-I-26A.

US Department of Commerce. Bureau of Census. Various years. *Census of manufactures, industry series: wooden containers and miscellaneous wood products.* Washington: USGPO document no. MC82-I-24C.

US Department of Labor. 1979. *Employment and earnings, 1909-78.* Washington: US Department of Labor Bulletin 1312-11.

US Department of Labor. 1983. *Supplement: employment and earnings.* Washington: US Department of Labor.

US Government Printing Office. 1982. *Economic report of the president*. Washington: USGPO.

Wallace, T.D., and D.H. Newman. 1986. Measurement of ownership effects on forest productivity: North Carolina 1974-1984. *Canadian Journal of Forest Research* 16(4):733-738.

Wallace, T.D., and J.L. Silver. 1984. Public cost sharing and production in nonindustrial private forests: the case of FIP in coastal Georgia. Pp. 165-180. In *Nonindustrial private forests: a review of economic and policy studies*, ed. J.P. Royer and C.D. Risbrudt. Durham, NC: Duke University School of Forestry and Environmental Studies.

Warren, W.G. 1990. Some novel statistical analyses relevant to the reported growth decline in pine species in the southeast. *Forest Science* 36(2):448-463.

Welch, F. 1971. Formal education and the distributive effects of agricultural research and extension. Pp. 183-192. In *Resource allocation in agricultural research*, ed. W.L. Fishel. Minneapolis: University of Minnesota Press.

Westgate, R.A. 1986. The economics of containerized forest tree seedling research in the United States. *Canadian Journal of Forest Research* 16:1007-1012.

Williams, C.G., and J.E. desteiguer. 1990. Value of production orchards based on two cycles of breeding and testing. *Forest Science* 36(1):156-168.

INDEX

Acreage index, 128
Adams, D. M., 30, 87
Africa, 227
Aggregate production, 200
Alig, R., 150
Amazon, 224
American Plywood Association, 61
American Wood Preservers Association, 102
Annual productivity change, 146
Arrow, K. J., 53
Asia, 227
Auxiliary torque, 70

Band-saw, 30
Benefit-cost (BC) ratio, 88, 89, 96, 174
Bengston, D., 20, 33, 35, 36, 37, 40, 57, 64, 75, 155
Bengston-Griliches approach, 75
Berck, P., 30
Best-open-face (BOF) technology, 82, 87, 88, 113
Biltmore Forestry School, 194
Binswanger, H., 224
Biological diversity, 197, 220, 226
Biological research, 32, 225
Biological variables, 32
Birdsey, R., 150, 151
Bogalusa, Louisiana, 154, 156, 162, 164, 165, 169, 170, 171, 212, 213
Boyce, S. G., 151
Boyd, R. G., 86, 184, 224, 227
Brannlund, R., 151
Brumm, H. J., 25
Buckman, R. E., 225
Buongiorno, J., 68, 84, 94, 96
Business Week, 42

Callaham, R. Z., 21, 32, 36, 69
Campbell, R. C., 167
Canada, 221, 222, 225, 226
Capacity expansion, 94
Capital, opportunity costs of, 7, 155
Capital gains taxation, 227
Capper Report, 176
Charcoal research, 223
Chip-n-saw, 31
Chromate copper arsenate treatment, 100
Clarke-McNary Act, 176
Climate change, 197, 220, 225
Cline, P. L., 24
Coastal North Carolina, 154, 156, 166, 169, 171, 179, 212, 213
Cobb-Douglas production function, 40, 42, 44, 57, 65, 67, 118, 119
Collinearity, 11, 57, 65
Commercial forest plantations, 225
Congo, 224
Conservation Reserve Program, 157
Consultative Group of International Agriculture Research, 227
Consumer benefits, 210
Consumers, 205
  of pulpwood, 146
  of sawmills, 15, 30, 31
  of solidwood, 146
  of stumpage, 148
Consumers' surplus, 6, 9, 21, 25, 33, 35, 36, 47, 49, 52, 146
  methodology, 5, 7, 18, 146
Cooperatives, industry and university, 16, 183, 184
Corn hybridization, 25
Cost cohort, 215
Cost of capital, 94, 96

Cost of capital (*continued*)
  coefficients, 83, 93, 102, 143, 144
  data, 63, 83, 93, 102, 109
  user, 63, 84, 94, 96, 102, 109, 138, 139
Cost-sharing programs, 217
Creosote, 100

Davis, L. S., 36, 73, 225
Debarking, mechanical, 82
Demand elasticities
  price, 67, 68, 76, 202
  pulpwood, 94, 96, 147
  sawmills, 87
  softwood plywood, 68
  solidwood, 140–142, 200
  wood preservatives, 104
Demand function, 66, 86, 93, 94, 103,
  110, 149, 204
Denison, E., 24, 196, 199
Department of Agriculture, U.S., 13
Department of Commerce, 137
Diffusion, 21
Dissemination costs, 69, 70
Distribution, 154, 155, 165, 170
  of technical change, 153
  of timber benefits, 164
Douglas-fir, 61
Drainage technology, 154
Dual approach, 18, 34, 37, 40, 59, 92, 99,
  200, 204, 205
Duke University, 179
Durbin's *h*, 66, 103, 142, 201
Durbin-Watson test, 67, 84, 95, 96, 140,
  142

Economic development, 197
Economic growth, 23, 196, 213
Economic rent, 6
Education, 24, 185
Efficiency gains, 213, 226
Employment, 160, 161, 164, 167, 172, 213
Endangered species, 220
Energy concerns, 219
Environmental concerns, 219
Environmental Protection Agency, 172
Environmental residuals, 198, 199, 205
Evenson, R. E., 57
Exports, 61
Extension, 12, 150, 173, 184, 199, 213,
  225

Factor costs, 35
Farm forestry, 225
Fertilizer, 19
Final good prices, 94, 138
Fire control, 31
Firm size, 18
Fleisher, H. O., 61
Flick, W. A., 162, 172
Forest, old growth, 228
Forest experiment stations, 124, 138, 151,
  176, 186, 192, 193, 213
Forest inventory, 17, 118, 119, 120
Forestland, 221
Forest ownership types, 115, 118
Forest productivity change, 154
Forest products industries, 13, 41, 81,
  113, 131, 196, 220
Forest Products Laboratory (FPL), 12,
  16, 23, 41, 58, 63, 65, 69, 70, 76, 84,
  87, 88, 92, 96, 99, 100–103, 106, 107,
  110, 113, 201, 205, 206, 219, 227
Forest products research, 4, 5, 20, 60, 75,
  100, 200, 219, 225
  cost-reducing, 81
  in developed countries, 222
  in developing countries, 223
  product-altering, 81
Forest research, 145, 147, 155, 161, 197,
  219
  in developing countries, 197, 221, 223
Forestry extension, 150, 173, 184, 199,
  213, 225
Forestry sectors, 154, 213, 217
Forest Service, 3, 33, 36, 70, 72, 121, 124,
  137, 153, 156, 162, 170, 176, 178,
  183, 189, 190, 192, 212, 213, 215, 220
  research budget, 179, 215
Forest survey, 124, 138, 178
Frontier model of economic development,
  28, 217
Fuelwood, 152

Gaylord Container, 162
Geissler, P. J., 68
General equilibrium, computable, 172
Genetic improvement, 183
Georgia, 153, 156, 157, 159, 162, 165,
  169, 170, 171, 212, 213
Gilless, J. K., 94, 96

Gluing techniques, 201
Grabowski, H. G., 53
Grayson, A. J., 21
Great Southern Lumber Company, 162
Greber, B. J., 131, 152
Gregersen, H., 33, 155
Griliches, Z., 25, 33, 36, 40, 46, 57, 68,
    70, 75
Gross benefit, 202, 213
Grossman, E. S., 132
Gross national product, 23, 63
Gross social benefit, 202
Guthrie, J. A., 94

Hallock, Hiram, 88
Harlicek, J., 72
Hay, D. A., 53
Haygreen, J., 90
Haynes, R. W., 30, 87
Hemphill, J. M., 25
Herbicide, 23
Hertford, R., 146, 148, 210
Hodges, D., 180, 185
Home improvement, 82, 86
Homogeneity, 119, 123, 129
Hotelling's lemma, 43, 44
Hughes, J. H., 167
Human capital, 227

IMPLAN, 153, 154, 156, 157, 159, 162,
    164, 172, 212
Implementation costs, 69, 70, 205
Industrial research budget, 183
Industries
    construction, 63, 82
    forest products, 13, 41, 81–113, 131,
        196
    home improvement, 82, 86
    independent logging, 156, 165
    in developing countries, 223
    lumber, 13, 31, 67, 117, 156, 228
    other wood products, 156
    particleboard, 57, 64
    structural, 35, 77, 219
    planing mill, 32
    plywood, 156, 159, 160, 161, 164, 165,
        167
    pulp and paper, 13, 32, 57, 94, 138,
        140, 149, 154, 156, 157, 159, 160,

        161, 164, 165, 167
    pulpmills, 57, 81
    sawmills, 81, 82, 156, 159, 160, 161,
        164, 167, 196, 200
    SIC code, 18, 20, 53, 54, 138, 139, 228
    softwood plywood, 16, 18, 34, 35, 36,
        40, 41, 53, 58, 59–80, 88, 109, 110,
        196, 200
    solidwood, 138, 154
    southern pine, 18, 19, 32, 114–132, 197
    timber growth and management, 17,
        196, 200
    wood preservatives, 81, 99, 197
    woodpulp, 92, 197, 200
Input
    capital, 11, 14, 23, 120, 133, 134, 193,
        209, 216
    labor, 14, 23, 120, 133, 193, 209, 216
    land, 23, 120, 193, 216
    multiple, 10
    multiple research, 11, 51
    quality, 10, 11, 21
    raw material, 134
    R&D, 51, 61, 174, 199
Input-output analysis (I-O),153–157, 167,
    170, 212, 213
Intermountain West, 28
Internal rate of return (IRR), 7, 50, 189,
    190
    public research, 44, 190
    sawmills, 88–90
    softwood plywood, 76
    timber research, 193
    woodpulp, 96, 98, 99
International Union of Forest Research
    Organizations (IUFRO), 4
Interregional impacts, 155
Inventory coefficients, 210
Inventory elasticity, 141, 142, 145
Inventory of timber, 131, 133, 135, 136,
    221

Japan, 222, 223, 226, 229
Johansson, P. O., 151, 228
Josephson, H. R., 194

k values, 146, 147
Kendrick, J. W., 132
Knight, H., 150, 151
Knutson, M., 24, 196, 199

Laarman, J., 229
Labor, 11, 28, 65, 82, 94, 118, 154
    coefficients, 103, 143
Lags, 65, 74, 75, 84, 103, 199, 208, 218
    adjustments, 10
    inverted V, 42, 58
    Koyck, 42, 45, 56, 80, 202
    production, 135
    research and implementation, 12, 40,
        41, 45, 48, 60, 61, 77, 84, 89, 93,
        109, 204, 208
    research-productivity lags, 175, 190,
        192, 193, 195, 204, 208, 215, 219
Lake States, 28, 228
Land management
    mixed oak-pine, 120, 128, 129, 130, 132
    natural pine, 116, 120, 128–130, 132
    plantation, 29, 30, 116, 120, 128–130,
        132, 157, 166
Land ownership, 208
    forest industry, 28, 129, 155, 162, 166,
        208, 210, 218
    NIPF, 12, 123, 125, 129, 130, 155, 157,
        162, 166, 208, 210, 217
    public, 123, 129, 132, 155, 162, 208
Lathes, 61
Latin America, 227
Lewandrowski, J. K., 87
Lewis, D., 88
Lofgren, K. G., 151, 228
Logging, in developing countries, 169
Lu, H-C., 84
Lutz, J. F., 61, 65
Lyon, K. S., 29, 30, 222

McIntire-Stennis Act, 179
McIntire-Stennis funds, 179, 180
McKillop, W., 68
Mairesse, J., 57, 68
Maloc, B., 194
Management-ownership index, 119, 120,
    125
Mansfield, E., 33, 53, 57
Marginal internal rate of return (MIRR),
    35, 45–47, 88–90, 203, 204
    private, 199
    softwood plywood, 70, 72, 74
    wood preservatives, 106, 107
    woodpulp, 96, 99
Marginal lands, 217

Marginal social returns, 206, 207
Marginal value product of research, 9, 44,
    61
Market concentration, 219
Market power, 207
Markets
    monopsonistic, 136
    oligopsonistic, 136, 155
Mergen, F., 227
Morris, D. J., 53
Multicollinearity, 67, 129
Multipliers, 88, 155, 172
    employment, 155
    Type I, 157, 162, 167, 172
    Type III, 157, 162, 167, 172

National Forest Management Act, 156
National Income and Products Account,
    156
National Science Foundation, 58
Nepal, 224
Nerlove, M., 53
Net economic benefit (NEB), 34, 35, 51,
    68, 70, 73, 98, 100
Net present value (NPV), 76, 77, 88, 89,
    96, 99, 189, 190, 193, 202
Net research benefits, 198
Net social gains, 21, 107, 111, 196, 200
New capital equipment, 201
New England, 28
Non-industrial private (NIPF) timber
    suppliers, 116, 120, 123, 125, 129,
    130, 150, 185
Non-linear two-stage least squares, 201
North America, 222, 229
North Carolina, 154, 156, 166, 169, 170,
    171, 212, 213
North Carolina State University, 183
Northeast, 228
Norton, G. W., 36
Nursery management, 228

Office of Management and Budget, 3, 72,
    189
Office of Science and Technology, 3
Office of Technology Assessment, 13
Otto, D. M., 72, 73
Output elasticity
    capital, 118, 130, 132
    of government R&D, 66, 67, 80

labor, 76, 80, 87, 202
of private R&D, 68, 202
of public R&D, 36, 37, 76, 95, 205, 206
Ownership, 118, 130, 132

Pacific Northwest, 35, 75, 218, 228
Pakes, A., 41, 57, 65
Papermills, 94
Particleboard, 35, 57, 64. *See also* Structural particleboard
Patents, 16, 227
Peterson, W. L., 46, 70
Philippines, 224, 229
Pine
  loblolly, 117
  longleaf, 116
  natural stands, 120, 128, 132
  plantations, 120, 157, 166, 167
  shortleaf, 116
Plywood production, 16, 137
Policy spillovers, 224, 225, 228
Political intervention, 206
Pollution abatement research, 92
Powered back-up roll, 16
Process innovation, 23, 220
Producer benefits, 150, 210
Producers, 205
  of pulpwood, 148, 149
  of stumpage, 148, 149
Producers' surplus, 6, 7, 9, 21, 25, 34, 39, 47, 50, 52, 146
  methodology, 5, 7, 18
  stumpage, 146
  wood preservatives, 105
Product innovation, 23
Production function, 5, 7, 8, 9, 10, 18, 21, 25, 34, 37, 40, 51, 91, 110, 198, 200, 209
  Cobb-Douglas, 40, 42, 44, 57, 65, 118, 119
Production technology, 13
Productivity shifts, 122, 128–130, 146, 154, 187
Product quality, 198, 199
Proprietary benefits, 207
Public policy, 198, 219
Public research, 4, 16, 20, 39, 59, 84, 87, 92, 96, 98, 103, 170, 196, 200, 207
  coefficient, 103
Public welfare, 198

Pulpmills, 57, 149
Pulpwood, 92, 116, 117, 135, 136, 145–147, 149, 152, 209, 210, 213, 228
  demand equation, 141
  price, 137
  price coefficient, 142
  stumpage, 137, 140

Rate of return, 7, 44, 50, 88
Real interest rate, 67
Regional economies, 213
Regulations, 199, 227
Relative prices, 228
Renewable Resources Planning Act, 3
Research. *See also* Research, areas of
  average returns to, 200, 205–207
  benefits, 59, 145, 147, 198, 206, 210
  budgets, 179, 183, 197, 215
  cost accounting, 10
  cost reducing, 81
  costs, private, 40
  elasticities, 202
  evaluation, 46, 197, 198
    consumer/producer surplus method, 6, 7, 9, 18, 21, 25
    ex post and *ex ante* method, 5, 18, 19, 20, 221
    index number method, 7, 25
    production function method, 18
    public welfare, 198
  federal, 175
  industrial, 183
  investments, 39, 200
    private, 39, 44, 199, 200
    public, 14, 39, 44, 59, 77, 106, 200, 204, 205
    rationale, 14
    rules-of-thumb, 53
    social gain, 14
    softwood plywood, 76
    time horizon, 15
  production, 110
  public. *See* Public research
Research, areas of
  American forestry, 219
  biological factors, 32, 225
  charcoal, 223
  fertilizer, 19
  forest products, 60, 75, 81, 200, 219, 225

Research, areas of (*continued*)
  in developed countries, 222
  in developing countries, 223
  growth and yield, 19, 175
  international forestry, 221
  lumber, 37
  pollution abatement, 92
  sawmill, 15, 18, 21, 30, 81, 82–91, 228
  softwood plywood, 16, 18, 40, 59–80,
    90, 147, 202–204, 207, 219
  southern pine, 132, 133, 153, 165, 169,
    170, 171, 174, 175, 178, 183, 190,
    193, 209, 213, 218
  timber growth and management, 5, 17,
    18, 19, 176, 197, 200, 208, 215,
    218, 221, 226
  in developed countries, 222
  in developing countries, 225
  weed control, 216, 228
  wood alcohol, 92
  wood preservatives, 18, 81, 99–112, 207
  woodpulp, 18, 81, 91–99, 110, 207
Research and development (R&D), 57
  coefficients, 95, 103
  costs of, 12, 25, 35, 36, 37, 39, 40, 63,
    77, 173, 174, 192, 199
  elasticity, 48, 66
  rationale for public role, 36, 73, 84, 87,
    92, 105, 219
  revenues and private R&D expenditures,
    18, 20, 36, 67, 69, 70, 80, 93
  rules-of-thumb, 53
  sawmills, 84
  softwood plywood, 64, 68, 70, 219
  wood preservatives, 105
  woodpulp, 92, 93
R&D expenditures by, 37, 38, 49, 105,
    175, 199, 213
  educational institutions, 175
  federal government, 12, 13, 49, 175
  forest industry, 34, 35, 41, 54, 64, 175,
    180
  forestry extension, 175, 213
  Forest Service, 213
    experiment stations, 175, 180, 183,
      185, 186. *See also* Forest experi-
      ment stations
    input, 51
    private sector, 20, 33, 36, 40–42, 51,
      53, 83, 112, 173, 202, 204, 205

  public sector, 12, 18, 20, 36, 41, 51,
    105, 161, 173, 200, 201, 202
  state, 12, 175, 179
  southern pine, 187, 191
  university/industry, 16, 173, 175, 179,
    180, 183, 186
Residential construction, 63
Residual approach, 9, 10
Risbrudt, C., 21
Robinson, V. L., 87, 131
Rockefeller Foundation, 3, 21
Rockel, M. L., 68
Rocky Mountains, 28, 218, 228
Roundwood, 217, 220
Royer, J., 184
Ruttan, Vernon, 5, 25, 37, 194, 208

Sallyards, M., 113
Sawmills, 30, 31, 32, 81, 82, 84, 152, 169,
    196, 200
  applications, 108
  benefit-cost ratio, 88–89
  best-open-face (BOF), 82, 87, 88
  capacity, 82
  cost of capital, 83, 84, 109
  internal rate of return (IRR), 88–90
  labor costs, 82
  labor productivity, 82
  marginal internal rate of return
    (MIRR), 88–90
  mechanical debarking, 82
  net present value (NPV), 88, 89
  private R&D expenditures, 83
  public R&D, 84, 87
  research, 90, 204, 228
  scientist months (government), 83
  supply, 83, 84
  value of marginal product (VMP), 88,
    89
  wages, 83, 84
Schaffer, W. A., 172
Schmitz, A., 146, 148, 210
Schumpeter, J., 111
Scientist time, 47, 69, 83, 93, 102
Sedjo, R., 29, 30, 222
Seed stock, 225
Serial correlation, 201
Shankerman, M., 41, 57, 65
Silver, J. L., 118
Silvicultural management, 217

Simultaneity bias, 57
Site quality, 119, 124, 125, 126
Social discount rate, 204
Social efficiency gains, 223, 226
Society of American Foresters, 4
Softwood, 135
  productivity, 116
  removal, 125, 166
Softwood plywood industry, 4, 16, 18, 34,
    35, 40, 41, 53, 58, 59–80, 109, 115,
    152, 196, 200, 202–204, 208
  cost of capital, 63
  dissemination costs, 69
  exports, 61
  implementation costs, 69
  lag, 84
  marginal internal rate of return
    (MIRR), 70
  net economic benefit, 70
  plants, 62
  price, 75
  research, 74, 77, 90, 219
    investments, 76
    returns, 73, 75, 207
  R&D expenditures, 64, 69, 70, 77
  residential construction, 63
  social discount rates, 72
  supply and demand, 61, 65, 78
  technology, 61
  value of marginal product (VMP), 70,
    71
  wages, 63, 64, 67, 75
Solidwood, 115, 116, 135, 137, 142, 145–
    148, 152, 209, 210, 213, 228
Solow, Robert, 24, 196, 199
South, 28, 32, 35, 218, 226, 228
Southern pine, 18, 19, 30, 114, 133, 152,
    153, 215, 216, 218, 220, 228
  growth and management, 186, 197
  inventory, 116, 118, 131
  plantation, 167
  research, 132, 165, 169–171, 174, 175,
    178, 183, 184, 190, 193, 209, 213,
    218, 221
  R&D, 173, 184, 187, 191
Southern pine industry, 115, 116, 193
  acreage index, 128
  capital, 103, 118
  forest ownership types, 114
  growth and management, 131

  homogeneity, 119, 123, 129
  inventory, 116, 119, 124, 131, 133, 135,
    136, 138, 209
  labor, 103, 118, 142
  loblolly, 117
  longleaf, 116
  lumber, 117, 137
  management-ownership index, 119, 120,
    125
  mixed oak-pine stands, 120, 128–130
  natural stands, 116, 120, 128, 129, 130
  NIPF, 116, 129
  plantations, 116, 120, 128, 129, 130
  prices, 114
  productivity, 128
  pulpwood, 116
  rotation, 209
  shortleaf, 116
  site quality, 119, 124, 126
  softwood productivity, 116, 133
  solidwood, 115, 116, 135
  stocking, 119, 124, 126
  timber management, 118, 131
  tree diameter, 119, 124, 126
Southern Plywood Association, 137
South's Fourth Forest, The, 125, 150, 151,
    185, 186
Soviet Union, 221, 222, 225, 226, 229
Spelter, H., 44
Stier, J. C., 131
Stocking, 119, 124, 126
Structural particleboard (SPB), 23, 35, 36,
    77, 219
Stuart, T. W., 68
Stumpage demand, 133, 154
  pulpwood, 133, 140, 142
  solidwood, 133, 142
Stumpage inventory, 144
Stumpage markets, 137, 147
Stumpage prices, 29, 133, 137, 144, 194,
    217, 223
Stumpage supply, 133, 137, 209
  pulpwood, 117, 133, 135, 140
  solidwood, 133, 135, 142
Sullivan, J., 180, 185
Sulphate pulping process, 30
Supply elasticities
  price, 37, 48, 68, 75, 76, 94, 202, 205
  public R&D, 48, 202
  pulpwood, 147, 210

Supply elasticities (*continued*)
  R&D, 48, 66
  sawmills, 84, 87
  softwood plywood, 66, 68
  solidwood, 140–142, 200, 210
  wood preservatives, 99
  woodpulp, 95, 110
Supply function, 18, 37–39, 43, 54, 56, 61, 83, 93, 94, 101, 110, 149, 198, 200, 202, 204, 209

TAMM, 44
Technical change, 5, 10, 11, 12, 21, 23, 24, 28, 198, 216, 218, 220
  cost-reducing, 103
  in developed countries, 222
  disembodied, 58, 65, 66, 80, 101, 103, 104, 201, 209, 227, 228
  embodied, 199
  labor-saving bias, 30, 32
  process vs. product quality, 198
  R&D costs, 13, 110, 154, 192, 196
  softwood plywood, 201
  in timber production, 6, 7, 9, 32, 82, 115, 118–120, 122, 124, 131–133, 147, 153, 154
Technology assessment, 13, 14, 30, 32, 198
  energy, 14
  logging, 223
  softwood plywood, 61, 203
Tenure, 224, 225
Texas A&M, 183
Thailand, 224
Timber growth and management, 4, 19, 115, 118, 131, 135, 154, 196, 200
  cost cohort, 133
  inventory, 131, 133, 135, 136, 221
  research, 5, 17–19, 114, 118, 176, 190, 192, 193, 197, 200, 208, 215, 218, 226
Timber harvest and utilization, 28
Timber-Mart South, 152, 159
Timber prices, 114, 228
Timber production, 28, 155, 197
  capital, 155
  labor, 155
  land, 155
  managerial inputs, 155
Total economic benefit, 30, 47, 50
Total industrial output (TIO), 157, 162,
164, 169, 213
Tree improvement, 15, 19, 184
Truss frame housing, 22, 23, 220
Tweeten, L. G., 24, 196, 199

Ulrich, A. H., 93, 138
University of Georgia, 179

Wage
  coefficients, 67, 83, 93, 102
  data, 63, 64, 75, 78, 83
Wallace-Silver analysis, 118
Water Resources Council, 72
Weed control research, 216, 228
Weir, R., 194
West Coast, 28
Western Europe, 222, 226, 229
Weyerhaeuser, 166
Wharton Econometrics, 139
White, D. E., 131
Wisdom, H. W., 152
Wood fiber, 222
Wood preservatives, 197, 204
  chromate copper arsenate treatment, 100
  cost of capital, 102
  creosote, 100
  demand function, 103
  lags, 103
  marginal internal rate of return (MIRR), 106, 107
  net social gains, 107
  price coefficient, 101, 105
  railroad ties, 101
  research, 106, 207
  scientist months, 102
  supply function, 99, 101
  value of marginal product (VMP), 106, 107
  wages, 102
Woodpulp, 197, 200, 204, 207
  benefit-cost ratios, 96
  capacity expansion, 94
  cost of capital, 93, 94, 96
  demand function, 94
  dual, 92
  final good prices, 94
  internal rate of return (IRR), 96, 98, 99
  marginal internal rate of return (MIRR), 96, 99

net economic benefits, 98
net present value (NPV), 96, 99
papermill capacity, 94
pollution abatement, 92
private R&D, 93
public R&D, 92, 95, 96, 98
pulp production, 91
scientist months, 93
supply function, 93, 94, 110
value of marginal product (VMP), 96
wages, 93, 94, 95
wood alcohol, 92
Wood utilization, 4